Arctic Plants of Svalbard

Yoo Kyung Lee

Arctic Plants of Svalbard

What We Learn From the Green in the
Treeless White World

 Springer

Yoo Kyung Lee
Division of Polar Life Sciences
Korea Polar Research Institute
Incheon, Korea (Republic of)

ISBN 978-3-030-34562-4   ISBN 978-3-030-34560-0   (eBook)
https://doi.org/10.1007/978-3-030-34560-0

Cover illustration: Mountain Avens (Dryas octopetala) - Longyearbyen, Svalbard 5th July 2015.

This Springer imprint is published by the registered company Springer Nature Switzerland AG.
The registered company address is: Gewerbestrasse 11, 6330 Cham, Switzerland

Dedicated to my parents who have constantly trusted, encouraged, and supported me throughout my life

Preface

This book is about the Arctic tundra and the plants living there. Because the Arctic is so wide, the scope of the story is narrowed down to Svalbard. Though Svalbard is small, here we can see most of the characteristics of the Arctic tundra. Svalbard has a variety of tundra, including polar deserts, dry tundra, mesic tundra, and wet tundra. Similar plants living in Svalbard are also found in Greenland, Alaska, Siberia, and the Canadian High Arctic. So, this book shows the Arctic environment and tundra plants through Svalbard plants.

This book consists of five chapters. The first chapter explains where the Arctic is, and what the Arctic tundra is. Where does the Arctic begin and end? In fact, there is still no official definition of the Arctic. A certain lady who was born and raised in Tromsø, located at 69°N, said that she had never thought that Tromsø was the Arctic when she was young. To her, the Arctic was an area covered with white ice somewhere in the north. But now, not only Tromsø but also Rovaniemi, Murmansk, Arkhangelsk, etc., claim to be the Arctic city. The concept of the Arctic has been expanded over time.

The second chapter describes the Arctic vegetation. In general, the Arctic is divided into the High Arctic and the Low Arctic, but the subarctic is not included in the Arctic. The plants that are commonly observed in the High Arctic are different from those that are easily visible in the Low Arctic. Plants can be divided into shrubs, forbs, graminoids, etc., according to their life form. Knowing the life form of these plants helps understand the function of plants in the ecosystem. Scientists have classified the Arctic vegetation into five Arctic zones. If you know what life form plants dominate in each Arctic zone, you can understand the characteristics of Arctic vegetation at a glance.

The third chapter explains the type of Arctic tundra with representative plants. There is more than one kind of tundra, for instance, polar desert, dry heath tundra, mesic tundra, moist tussock tundra, wet sedge tundra, and shrubs tundra. Here, I introduce the characteristics of each tundra and the plants that represent each respective tundra.

The fourth chapter shows the hidden story of tundra plants through 10 representative plants in Svalbard. I introduce endangered species in Svalbard because they are the plants that we should know and protect. Next, I introduce plants without flowers such as mosses, liverworts, clubmoss, horsetail, and moonwort. Then, I move to the flowering plant. Through purple saxifrage, whitlow-grasses, and Arctic mouse-ear, we can understand how Arctic plants overcame the harsh environment. Alpine bistort, polar campion, and willows explain the mystery of plants in terms of developmental biology and genetics. Moss campion and mountain sorrel show the benefits of living together. The settlement process in the Arctic of mountain avens and buttercups reveals the migration history of Arctic plants. You will see that each plant has its own story.

The fifth chapter examines the process of how plants settled in the tundra after the Ice Age. It also describes how climate change will affect Arctic plants in the future. Chapters 1–3 contain general information of Arctic and tundra plants, but Chap. 4 contains specific examples. If you are interested in the story of Arctic plants, it would be good to start reading Chap. 4.

This book provides indexes to make it easier to find the common names and scientific names of the plants mentioned in this book. The list of vascular plants living in Svalbard is also listed at the end.

This book does not include the flower language of the plant, its use, and the traditional knowledge of Inuit. Instead, it contains results that scientists have observed, investigated, and analyzed over the past 100 years. I tried to write this book as simple as possible so that even high school students can understand it. If there is something difficult to understand, it is because I have not been able to interpret the science into general language. Through this book, I hope to strike an interest in you about the importance of conserving Arctic plants which are disappearing.

Incheon, Korea   Yoo Kyung Lee

Acknowledgments

This book was conceived through an e-mail. An associate editor, Emmy Lee, at Springer Nature Korea Limited was looking for future authors who were planning to publish books. This kind of e-mail was usually thrown directly into the trash, but then this caught my interest because I was going to go to Tromsø, Norway, for 6 months.

My workplace, the Korea Polar Research Institute (KOPRI), was operating the KOPRI-NPI (Norwegian Polar Institute) Cooperative Polar Research Centre to enhance research cooperation with the NPI. The office of the center was located in Tromsø. Tromsø is a small island, which is a couple of hours of flight away from Oslo. Tromsø, nicknamed "the Paris of the Arctic," has been a place where many tourists flock all year round to see the Aurora and the fjord. My mission was to find eye-catching research cooperative items, but it did not seem to take much time. Looking for work during the time, I visited Springer Nature's Korean office with a vague idea of writing a book about Arctic plants.

I wanted to write a science book introducing Arctic plants to the public: simple, short, and easy to understand. Emmy Lee was interested in my plan and sent me a book proposal form. I carefully wrote a proposal, which was then delivered to Christina Eckey, executive editor of plant sciences, Dr. Éva Lørinczi, associate editor, and Abinay Subramaniam, project coordinator, in turn. Two reviewers encouraged me to add explanations of Svalbard and cryptogamic plants. I am grateful to these people who have helped me write this book.

In 2003, when I first visited Svalbard, I did not notice most of the flowers at that time because I was obsessed with microbial sampling. I cultivated a variety of bacteria that lived in Svalbard soil and seawater, and I have found a new bacterium and named it *Dasania*. *Dasania* was named in honor of Dasan, a Korean scientist Jung Yak-Yong. He was one of the greatest scholars who wrote highly influential books about philosophy, science, and theories of government. He was so interested in plants that he even wrote books about crops. Of course, he did not have time to write a book when he was on the political scene. When he lost political power and went to exile on the outskirts of Korea, he could find time to write many books. In any case, scholars need time to concentrate on their research. I am also grateful to KOPRI for giving me this precious gift, time.

I would like to thank the authors of many papers whether they were cited in this book or not. Thanks to their expedition through the Arctic tundra, observing, questioning, investigating, and keeping records, I was able to write line by line in this book. I would like to thank KOPRI librarians, Hyunyi Park, and Miyeon Kim for finding numerous papers. I am also grateful to the Head librarian of NPI, Ivar Stokkeland. He found books I could not find in Korea, and sometimes he borrowed them from other libraries for me.

Most of the pictures were taken during the Arctic expedition supported by the Ministry of Science and ICT and by the National Research Foundation of the Republic of Korea (2016M1A5A1901769, KOPRI-PN20081). I cannot skip thanking Prof. Skip Walker, Dr. Ji Young Jung, Youngsim Hwang, Blueshade, Robert Rohde, Oona Räisänen, Leland McInnes, Paataliputr, the Alaska Geobotany Center, CAFF (Conservation of Arctic Flora and Fauna), Smithsonian Institution, and Wikimedia for letting me use their photos or figures.

I am also grateful for Google Translator and Naver Papago. Without them, I would not have tried to write a book in English. I am thankful to Somang Jeong and Jane Lee who have refined the English expression of this book. Thanks to their corrections, the sentences have become much more natural.

Finally, I thank my family. My mother-in-law usually serves meals for family members. My husband has always respected my choice to stay in Tromsø or to have frequent business trips. Hannah and David have been the driving force behind my life. Without their support and help, I might have given up on the way of being a scientist.

Prologue

Is the Color of the Arctic White?

"I am a scientist conducting research in the Arctic."

"Oh, then, you must have seen polar bears on white ice!"

An endless stretch of snowy ice with polar bears wandering about, skipping across sea ice... When I say, "the Arctic," the first thing that comes to most people's mind is the image of a white land covered by snowy ice and roaming polar bears on it. But is, in fact, the color of the Arctic white?

In the summer of 2016, I went to Alaska with Dr. Youngwook Kim who studies ecosystems by analyzing satellite data. He accompanied me in doing field research work to investigate chlorophyll fluorescence in the Arctic observed by satellite data.

He said, "The amount of chlorophyll fluorescence in Alaska was found to be too high. It is a value, in which the surface should be entirely covered by vegetation. Yet there are not supposed to be trees in the Arctic; there cannot be so many plants."

After going to the Alaskan field, however, his concept of Arctic vegetation changed. He admitted, "I did not know there were so many plants in the Arctic. This coverage of plants explains why the amount of chlorophyll was so high."

That is to say, the color of wet tundra in the Arctic is green (Fig. 1). There is also dry tundra. There are even desert tundra in the Arctic, and these dry areas are gray-brown. In fact, there are various kinds of vegetation in the Arctic, with blossoms and deciduous leaves, yellow, red, orange, and white mixed with green. That being said, the Arctic is not just white; it is multi-colored.

Fig. 1 "The Green Arctic": a field full of vegetation on the way to Teller, Alaska

Contents

Chapter 1
Arctic Tundra: Where There Are No Trees

Where Is the Arctic?

Antarctica has an official definition but the Arctic has not. The Antarctic Treaty defines the Antarctica as all of the land and ice shelves south of 60°S latitude; thus, the Antarctic begins from the baseline of 60°S latitude. Then, where does the Arctic begin? In general, there are two types of definitions of the Arctic: the geological and the ecological. From a geological perspective, the Arctic is the area north of the Arctic Circle, the baseline of the midnight sun (Fig. 1.1). The midnight sun is a phenomenon that occurs because the Earth's axis tilted. Although the Arctic Circle, which serves as the baseline of the Arctic, is known to be 66.33°, it actually changes little by little every year because the Earth's axis changes little by little. For example, the Arctic Circle, which was 66°33′41″ North in 2001, changed to 66°33′54″ in 2018.

From an ecological point of view, the Arctic is the region north of the isotherm where the average temperature in July is less than 10 °C. This isotherm is similar to the tree line above which trees cannot grow. In 1951, Nicholas Polunin, a botanist and one of the world's foremost activist of environmental conservation, suggested this Arctic concept. Polunin, whose father had been a Russian-born forester, was born in England and studied at Yale and Oxford University. Since he had joined an Arctic expedition in 1931, he had taken part in botanical explorations to many places including Svalbard, Greenland, Iceland, Lapland, Labrador, and various islands of the Canadian Eastern Arctic. He proposed a set of criteria for delimiting the Arctic, following 20 years of field experience studying the Arctic plants: (1) a line 80 km north of the northern limit of coniferous forest; (2) north of the northern limit of trees of 2~8 m in height; or (3) north of the 10 °C isotherm of the warmest month[1]. His suggestion became the basis for the ecological definition of the Arctic (Fig. 1.2).

---

[1]The original criterion is north of the northern Nordenskiöld line, which is determined by the formula $V = 9-0.1K$ (V is the mean of the warmest month, and K the mean of the coldest month).

© Springer Nature Switzerland AG 2020
Y. K. Lee, *Arctic Plants of Svalbard*, https://doi.org/10.1007/978-3-030-34560-0_1

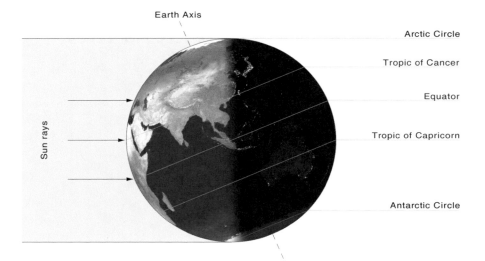

Fig. 1.1 Arctic Circle. In the summer, the Earth's axis tilts toward the sun. In the northern area of the Arctic Circle, the sun remains above the horizon for 24 hours during this specific season (© Blueshade, available via license: CC BY 2.0)

A region where trees cannot grow due to low temperature and short growth period is called "tundra," found in alpine as well as polar regions. Therefore, we can define the Arctic as the Arctic Ocean and the tundra region above 50° north. The tree line is the southern boundary of the Arctic (Fig. 1.3). In fact, it is not easy to distinguish regions of Arctic tundra from the subarctic regions. Because the tree line is not an actual line, we cannot delineate clearly from where the Arctic starts. Tree line is the transition area, which stretches across a narrow 10~50 km band in North America, and a broader zone in Russia. In Finnmark and Kola Peninsula, the Arctic region accounts for the narrow treeless zone north of the northernmost downy birch forests. Here, trees grow well in riparian habitats due to permafrost thawing by river water. Treeless regions are scattered across the mountain boreal area. Moreover, it is not impossible to find a tree line on the ocean. Nevertheless, we can say the Arctic is Arctic Ocean and treeless Arctic tundra.

In the ecological Arctic, the latitude varies by region. The ecological Arctic line of Siberia is farther north from the Arctic Circle, whereas it is lower for the Canadian Arctic. Trees grow up to a latitude of 70° north in Norway but they grow around 54° north in Hudson Bay, Canada. If the latitudes are the same, the amount of energy coming from the sun should also be the same. Why then do these differences arise?

The most basic element to determine the climate is solar energy, but climate depends on various other factors (Fig. 1.4). The energy delivered by ocean currents

---

Polunin mentioned, however, it was an empirical line, which seems an improvement at least on the 10 °C isotherm.

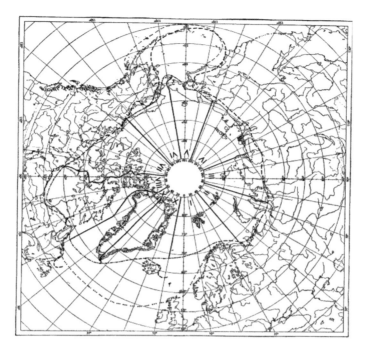

Fig. 1.2 The ecological definition of the Arctic. The northern limit of coniferous trees growth (dots) and isotherm line (long dashes). This map originated from a previous paper by Polunin (1951) (© The British Ecological Society)

is also one of the key factors that influence the climate. In Norway, the warm Gulf Stream passes by the coast of Norway and delivers warm energy. In contrast, eastern Canada is cooled from the transpolar current flowing down into Hudson Bay. Because of this, Norway is warmer than Canada at the same latitude. Thus, the ecological definition on the Arctic based on the average temperature or the vegetation better reflects the actual natural environment than the Arctic Circle connecting the same latitude.

Why No Trees in the Arctic?

Most people will think that the Arctic is too cold for trees to grow. To put it differently, most will assume that the annual temperature is important for tree growth. For trees to grow, however, the average summer temperature is more important than the average annual temperature. For instance, Yakutsk, which is located in the middle of Siberia, is much colder during the winter with the annual average temperature is lower than in Svalbard. In Yakutsk, however, there is an abundance of tall conifer trees, whereas there are no trees in Svalbard (Fig. 1.5). This

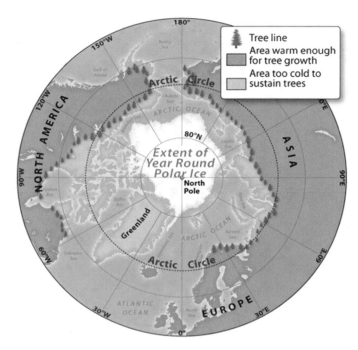

Fig. 1.3 The Arctic is defined by the 10 °C isotherm or the tree line as well as by the Arctic Circle (Created by The Map Factory. © Smithsonian Institution)

is due to the monthly average temperatures in June, July, and August, which are above 10 °C in Yakutsk, but below 10 °C in Svalbard. For the growth of trees, the atmospheric temperature during growing season (summer) is more important than the annual average temperature.

Alongside atmospheric temperature, other environmental factors are also important for tree growth. In Arctic tundra, the ground consists of frozen permafrost throughout the year. While permafrost is frozen soil by definition, a surface of the soil thaws in the summer. The thawing surface is called as the active layer, and plants take root and absorb nutrients in the active layer. As plants grow taller, they deepen their roots and hold themselves up. In the tundra, however, hard permafrost prevents the roots from reaching deep. Permafrost is a barrier to tundra plants. Tall plants with shallow roots collapse easily in the breeze of the Arctic. The depth of the active layer in Yakutsk of Siberia is about 2 m, but that in Svalbard is less than 1 m deep. The active layer depth is one of the factors that determine the boundaries of the tree distribution.

Poor nutrition in tundra soils is also a limiting factor for plant growth. Tundra plants generally grow in an environment that lacks nutrients, especially nitrogen and phosphorus. The roots of tundra plants cannot descend below the permafrost; therefore, space, where the plants can yield nutrients, is limited to the active layer.

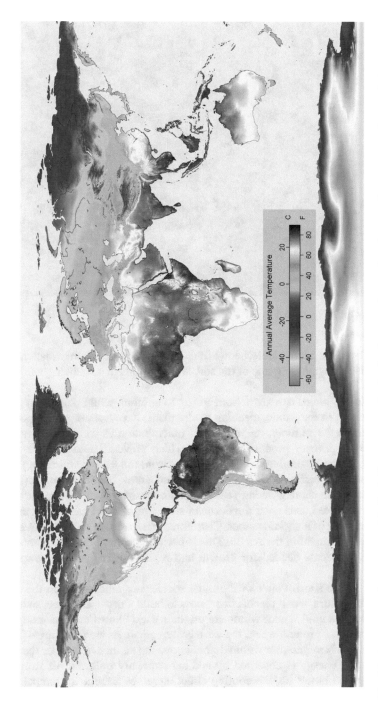

Fig. 1.4 Average annual temperature of the Earth. The temperature is related to the energy income from the sun and ocean streams. In general, the western part of the continents such as Alaska and Norway is warmer than the eastern part, i.e., the Russian Far East and Eastern Canada (© Robert A. Rohde/Berkeley Earth, available via license: CC BY 4.0)

Fig. 1.5 Treeless tundra in Svalbard

In addition, lower the temperature, slower the degradation of organic matter, and hence slower is the weathering rate of the soil, which results in a smaller quantity of nutrients.

In mid-latitudes, plants can grow from spring to autumn, while plants can grow only in the summer in the Arctic, meaning tundra plants have a short growth period. Moreover, even in the summer, the amount of precipitation as well as the temperature is low, thus primary production is low, which prevents tundra plants from growing large. At the beginning of the growing season, the land is still frozen, and toward the end, the available light is reduced and snow begins to accumulate. Therefore, during the short growing season, plants must break the dormant state, grow, bloom, bear seed, and enter the dormant state again. It not easy for plants to finish all the process in a short summer. Therefore, there are few annual plants and perennial plants dominant in the Arctic. This is because perennial tundra plants can accumulate the nutrients needed for growth and reproduction over the course of many years.

Tundra plants are short. Plants of the same species are much taller in the taiga region. In Arctic tundra, small plants allow snow to build a protective layer over the plant before dry and cold winter winds are on their way. Short tundra plants grow near the surface of the ground, where the warmest temperatures are maintained in the summer. Indeed, air temperature within 30 cm above the surface is warmer than the air above as solar energy is absorbed by and subsequently reflected off from the ground. If the wind blows up, the cooling effect increases and the air temperature

Fig. 1.6 Walking over dwarf shrubs in a tundra field (Courtesy of Dr. Ji Young Jung)

near a surface and at the far point may differ by 8 °C. Therefore, plants in the cold tundra grow close to the warmer surface as a strategy of survival.

In the forest, we must walk under trees. In tundra, however, we walk over dwarf plants (Fig. 1.6). Thus in Arctic tundra, we become little giants.

Bibliography

http://forces.si.edu/arctic/04_00_08.html
https://commons.wikimedia.org/wiki/File:Earth-lighting-summer-solstice_EN.png
Polunin N (1951) The real Arctic: suggestions for its delimitation, subdivision and characterization. J Ecol 39:308–315

Chapter 2
Arctic Is Not One

The High Arctic and the Low Arctic

Arctic is a huge area, and the Arctic environment is not uniform. They roughly divide Arctic into two regions: the High and the Low Arctic (Fig. 2.1). The High Arctic is the northern region of the Arctic with extremely cold winters and short cool summers. The ground of High Arctic is permafrost frozen to a depth of several hundred meters. Even though the transition between the Low and High Arctic is not very clearly delineated, the High Arctic differs from the Low Arctic in having very low-growing herbs and the absence of erect shrubs.

The delineation between the Low and High Arctic follow the 8 °C July isotherm in Canada, whereas it is closer to the 4 °C July isotherm in Eurasia. The growing season of the High Arctic is a shorter (2~2.5 months) than that of the Low Arctic (3~4 months). Summer is cooler in the High Arctic (July mean 2~8 °C) than in the Low Arctic (4~11 °C). The High Arctic has less precipitation than in the Low Arctic. Plant diversity and biomass are also low in the High Arctic. The soil-forming processes and decomposition is significantly slow in the High Arctic.

The High Arctic occupies approximately 35% of the Arctic tundra ice-free land area, and comprises polar desert, polar semidesert, and mires. Polar deserts and semideserts cover approximately 33% of the Arctic tundra, and mires have scattered in the High Arctic with only 2.4% of coverage.

Polar desert is a place that is arid and windswept. Scattered herbs (rooted poppy and smooth whitlow grass), rosette or cushions-forming species (purple saxifrage and early sandwort), and miniature grasses (alkali grass and ice grass) are the main inhabitants of the polar desert. Mosses and lichens are present only in traces.

Polar semidesert has more moist soils, of which typical species are prostrate shrubs (Arctic willow) and cushion plants (Arctic avens, mountain avens, moss campion, and saxifrages). They are commonly associated with herbs, drought-tolerant sedges (spike sedge and curly sedge), and grasses (Blue grass and Fescue). Mosses and lichens are more abundant.

Y. K. Lee, *Arctic Plants of Svalbard*, https://doi.org/10.1007/978-3-030-34560-0_2

Fig. 2.1 The High Arctic, Low Arctic, and subarctic boundaries (© Conservation of Arctic Flora and Fauna)

Mires and wet meadow are developed in moist areas such as snowbed and riparian. Arctic cottongrass, tall cottongrass, and water sedge are codominant species, and moss is a common companion. These wetland assemblages cover the High Arctic biomes.

In the Low Arctic, dwarf shrub, heath, tussock, sedge, and cottongrass make colorful vegetation. Dwarf birch, alder, and various species of willow are typical in the shrub tundra. Berry producing plants (blueberry, crowberry, and manzanita), Labrador tea, and Lapland rosebay are common in the heath field. Sedges, cottongrasses, and bulrushes are abundant in the mires of the Low Arctic. Hare's-tail cottongrass tussocks are important in surface hydrology in many Low Arctic ecosystems, but are absent in the High Arctic.

The Life Forms of the Arctic Plants

For better understanding of the Arctic vegetation, it would be good to know plant life forms such as tree, shrub, forb, and graminoid (Table 2.1). Arctic is a treeless place. A tree is a tall plant with single woody stem. White spruce, black spruce, American larch, balsam poplar, quaking aspen, and paper birch are representative trees that can be easily found at the border of tree line. Shrub is a multiple-stemmed woody plant, with small-to-medium size (Figs. 2.2, 2.3, and 2.4). Arctic willow, polar willow, green alder, American dwarf birch, Labrador tea, cowberry, crowberry, and moss bell heather are representative Arctic shrubs. Forb is an herbaceous flowering plant with broad leaves (Fig. 2.5). Most forbs are dicots, and both annual and perennial plants. Arctic poppy, alpine draba, and cloudberry are common Arctic forbs. Graminoid is an herbaceous plant with grass-like narrow or linear leaves (Fig. 2.6). Grasses, rushes, and sedges are representative graminoids. Moss is a small, soft, and flowerless plant growing in mats or clumps. Lichen is a symbiotic organism of filaments of multiple fungi and photosynthetic algae or cyanobacteria.

Five Bioclimatic Subzones of the Arctic

Early Arctic botanists and ecologists had proposed different types of zones in the Arctic based on climatic gradients, growth form of dominant plants, vegetation cover, geomorphology, hydrology, soil pH, soil moisture, snow accumulation, and so on. The name of zones was different from each other according to their research areas (Table 2.2). North American scientists studied in Alaska and Canadian High Arctic, Russians focused on Northern Siberia, and Norwegian researchers tried to compare Svalbard to adjacent Arctic areas. Russian and European researchers developed five zones, respectively. They combined their systems into a zonal subdivision, which was adopted by American researchers (Table 2.2).

In March 1992, two important meetings took place for Arctic vegetation study. Forty-four experts came together in Boulder, Colorado to join the International Workshop on Classification of Arctic Vegetation. They reviewed previous research and began the task of completing a global synthesis of Arctic vegetation. For the synthesis, they had to overcome the differences in language and scientific heritage among Arctic ecologists. As European, North Americans, and the Russians had their own traditions, the work began with an effort to understand their difference in taxonomic base, to revise synonyms for the same species, to compare the Arctic zonation, and to find the common mapping methods.

Independently of the Boulder workshop, the CAFF (Conservation of Arctic Flora and Fauna), one of the working groups of the Arctic Council, recognized the need to address the general lack of data on Arctic vegetation. Fifty-five scientists gathered in the historic village of Lakta near St. Petersburg, Russia. They agreed that they needed to review the status of Arctic vegetation mapping, and to develop a strategy

Table 2.1 Life forms of Arctic plants

Plant life forms (plant functional types)					Examples
Vascular	Woody	Erect shrubs (>0.1 m)	Low shrubs (0.4~2 m)	Deciduous low shrubs	Alder Dwarf birch Felt leaf willow Diamond leaf willow
			Dwarf shrubs (0.1~0.4 m)	Deciduous dwarf shrubs	Northern bilberry Lanate willow
				Evergreen dwarf shrubs	Ledum Crowberry Moss bell heather
		Prostrate dwarf shrubs (<0.1 m)	Deciduous prostrate dwarf shrubs		Arctic willow Polar willow Net-leaf willow
			Evergreen prostrate dwarf shrubs		Alpine bearberry Arctic bell-heather Lingonberry
	Non-woody (herbaceous)	Forbs	Cushion and rosette forbs		Saxifrage Draba Moss campion
			Other forbs		Lousewort Tundra milk vetch
		Graminoids	Grasses		Bluegrass Fescue
			Rushes		Wood-rush Common rush
			Sedges	Tussock sedges	Tussock cottonsedge Spruce muskeg sedge
				Non-tussock sedges	*Carex* spp. *Eriophorum* spp. *Kobresia* spp.
Nonvascular	Bryophyte	Sphagnoid moss			*Sphagnum*
		Non-sphagnoid mosses and Liverworts			Thread moss Wood-moss Bog groove-moss Fringewort

(continued)

Table 2.1 (continued)

Plant life forms (plant functional types)			Examples
	Lichen	Reindeer lichens	*Cladonia* (branching forms)
		Other lichens	Rim lichens Tile lichens Whiteworm lichens Dog lichens Rock tripe Iceland moss

Modified from Walker (2000)

Fig. 2.2 Low shrub with woody stems

for synthesizing the existing information into unified maps of Arctic vegetation. It was the first Circumpolar Arctic Vegetation Mapping (CAVM) Workshop. They drew out the basic objectives to produce (1) an internationally accepted geobotanical concept of Arctic vegetation distribution and zonation, (2) a photo quality, cloud- and snow-free, false-color infrared image, (3) a map of the relative vegetation greenness (i.e., biomass), and (4) a geobotanical database and derived maps of the circumpolar Arctic region.

Fig. 2.3 Deciduous dwarf shrub

Fig. 2.4 Prostrate dwarf shrub

There have been different classification systems of Arctic vegetation. Some researchers have described the vegetation based on the dominant structure of vascular plants in the North American Arctic. Dwarf-shrub heath, tussock tundra, sedge-moss meadows, and cryptogam-herb tundra are examples of this classification. Other researchers have recognized types of vegetation by the floristic composition and their relationship to the environment in Eurasian Arctic. Other researchers have categorized tundra plants based on plant growth form, life history, and physiological strategies. It took time to make a consensus of Arctic zonation.

Fig. 2.5 Cushion forb

Fig. 2.6 Rush, one of graminoids

After three more CVAM workshops, they published the first circumpolar vegetation map in 2002. They adopted five bioclimatic subzones A–E, in the Arctic (Table 2.3, Fig. 2.7). Subzone A is the coldest, and Subzone E, the warmest. The term "bioclimatic" means this zonation reflects climate as well as vegetation. Where plants grow and how much they grow depend on the long-term climate. Sometimes the growth form of plants responds to climate. Therefore, plants have

Table 2.2 Survey of Arctic zones

CAVM 2002 (Walker et al. 2002)	North America		Russia		Fennoscandia	
	Polunin (1951)	Daniels et al. (2000)	Yurtsev (1994)	Chernov and Matveyeva (1997)	Tuhkanen (1986)	Elvebakk (1999)
A (Herb subzone)	High Arctic	Arctic herb zone	High Arctic tundra	Polar desert	Inner polar zone / Outer polar zone	Arctic polar desert
B (Prostrate dwarf shrub subzone)	Middle Arctic	Northern Arctic dwarf shrub zone	Arctic tundra: Northern variant	Arctic tundra	Northern Arctic zone	Northern Arctic tundra
C (Hemi-prostrate dwarf shrub subzone)		Middle Arctic dwarf shrub zone	Arctic tundra: Southern variant	Typical tundra	Middle Arctic zone	Middle Arctic tundra
D (Erect dwarf shrub subzone)	Low Arctic	Southern Arctic dwarf shrub zone	Northern Hypoarctic tundra		Southern Arctic zone	Southern Arctic tundra
E (Low dwarf shrub subzone)		Arctic shrub zone	Southern Hypoarctic tundra	Southern (Shrub & tussock) tundra		Arctic shrub tundra

Modified from CAVM Team (2003)

been considered as living weather stations. The fundamental basis of the bioclimatic zonation was summer temperature. The subzones depend on mean July temperature and summer warmth index. When the mean July temperature is near 0 °C, the Arctic plants are at their metabolic limits. As the temperatures rise, plants' size and diversity increase.

The terrestrial area of the Arctic is 7.1 million km^2 (5.1 million km^2 are ice free). Excluding glaciers, Subzone A covers 0.1 million km^2 (2%), Subzone B covers 0.5 million km^2 (9%), Subzone C covers 1.2 million km^2 (23%), Subzone D covers 1.6 million km^2 (30%), and Subzone E covers 1.8 million km^2 (36%).

In Subzone A, mean July temperatures are less than 3 °C. Most parts of the land surfaces are barren, often with <5% cover of vascular plants. Subzone A appears to be very homogeneous at a circumpolar scale. Woody plants are absent. Lichens, bryophytes, and short forbs such as poppy, draba, saxifrage, and stitchwort scattered on the dry land. Common rush, foxtail, alkali grass, ice grass, and tundra grass are the main graminoids, but sedges are rare. There are no peat-producing mires. Subzone A is called as "herb subzone," because the dominant vascular plants are herbaceous species.

Table 2.3 Vegetation properties in each bioclimatic subzone

Subzone	Mean July temp. (°C)[a]	Summer warmth index (°C)[b]	Vertical structure of plant cover[c]	Horizontal structure of plant cover[c]
A (Herb subzone)	0~3	<6	Mostly barren. In favorable microsites, one lichen or moss layer <2 cm, very scattered vascular plants hardly exceeding the moss layer	<5% cover of vascular plants, up to 40% cover by mosses and lichens
B (Prostrate dwarf shrub subzone)	3~5	6~9	Two layers, moss layer 1–3 cm thick and herbaceous layer, 5–10 cm tall, prostrate dwarf shrubs <5 cm tall	5~25% cover of vascular plants, up to 60% cover by cryptogams
C (Hemi-prostrate dwarf shrub subzone)	5~7	9~12	Two layers, moss layer 3–5 cm thick and herbaceous layer 5–10 cm tall, prostrate and hemi-prostrate dwarf shrubs <15 cm tall	5~50% cover of vascular plants, open patchy vegetation
D (Erect dwarf shrub subzone)	7~9	12~20	Two layers, moss layer 5–10 cm thick and herbaceous and dwarf shrub layer 10–40 cm tall	50~80% cover of vascular plants, interrupted closed vegetation
E (Low dwarf shrub subzone)	9~12	20~35	2–3 layers, moss layer 5–10 cm thick, herbaceous/dwarf shrub layer 20–50 cm tall, sometimes with low-shrub layer to 80 cm	80~100% cover of vascular plants, closed canopy

Modified from CAVM Team (2003)
[a]Mean July temperatures based on Edlund (1990)
[b]Sum of mean monthly temperatures greater than 0 °C, modified from Young (1971)
[c]Vertical and horizontal vegetation structure based on Chernov and Matveyeva (1997)

In subzone B, mean July temperatures are about 3~5 °C. Woody plants first occur in this subzone, of which form is prostrate dwarf shrub. Therefore, "prostrate dwarf shrub subzone" is alternative name of subzone B. The representative prostrate dwarf shrubs are Arctic willow, polar willow, and mountain avens (Fig. 2.8). Draba, mouse-ear, poppy, saxifrage, and stitchwort are common forbs in this subzone. We can easily observe graminoids such as curly sedge, water sedge, alpine foxtail, dwarf hair-grass, and northern wood rush. Wood rush is an important component of mesic environment, and sedge and cotton grass often are dominant in wet areas including real mires.

Subzone C has more species diversity than Subzone B. Subzone C is characterized by a circumpolar occurrence of the hemi-prostrate dwarf shrub, Arctic bell-heather. This plant is prominent on large areas of acidic substrates, while rare on mesic alkaline surfaces. Legumes such as oxytrope and milkvetch are important in nonacidic regions. Dwarf fireweed is distinctive in creek sides (Fig. 2.9). Bellardi bog sedge, Bigelow's sedge, and curly sedge are common on upland surfaces, and

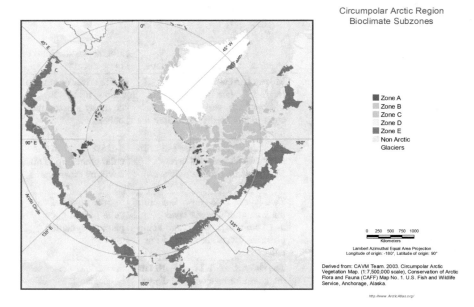

Fig. 2.7 Bioclimatic subzones in the Arctic [CAVM Team (2003), this map was modified from the previous reports by Yurtsev (1994) and Elvebakk et al. (1999)] (© The Alaska Geobotany Center)

Fig. 2.8 Mountain avens, a representative prostrate dwarf shrub of subzone B

Fig. 2.9 Dwarf fireweed in Subzone C

russet sedge and parallel sedge on calcareous substrates. Hoppner's sedge is more widespread in mires as well as on seashore.

Subzone D is located between the northern drier tundra on mineral soils and the southern relatively moist tundra with moss carpets and peaty soils. Peaty soils and fine-grained soils are often nonacidic, whereas soils in Subzone E are usually acidic. The moss layer, consisting primarily of aulacomnium moss, tomentypnum moss, wood-moss, and bog-moss, contributes to development of peaty and fine-grained soils. There is more regional variation than in Subzones B and C. Vascular plants generally cover about 50~80% of the surface. Dwarf birch is a common erect dwarf shrub (<40 cm tall) in this subzone. Mountain avens grows with prostrate shrubs such as alpine bearberry and pincushion plants. Arctic willow, net-leaf willow, and Arctic mountain avens are also common prostrate shrubs. Low shrubs (>40 cm tall) occur along streams. Richardson's willow and diamond-leaf willow are representative low shrubs. Felt-leaf willow can grow to tall shrub along rivers. Bigelow's sedge, membranous sedge, Bellardi bog sedge, and tall cottongrass are common graminoids. We can find tussock of hare's-tail cottongrass in Subzone D.

Subzone E is the warmest area in the Arctic tundra with mean July temperatures of 9~12 °C. Subzone E has taller shrubs and more cover of shrub than in Subzone D. Low shrubs of birches, willows, and alder are dominant vegetation in this subzone. Birch or willow thickets of 0.8~2 m tall occur in some moister areas, but the shrubs are shorter in areas with less snow cover. Tussock tundra is common in northern Alaska and eastern Siberia (Fig. 2.10). In some southern part of Subzone E, patches of open forest penetrate along riparian corridors. These forests consist of a

Fig. 2.10 Tussock tundra observed in Subzones D and E

variety of species of cottonwood, larch pine, spruce, and tree birches. This subzone has also been called the "low-shrub subzone".

In short, Subzone A has the characteristics of absence or only marginal presence of prostrate shrubs, and Subzone B with presence of prostrate shrubs but lack or only marginal presence of the dwarf shrub, Arctic bell-heather. Subzones C and D have dwarf shrubs, and Subzone E has low shrubs. This is a general concept, and sometimes a sharp zonation is not easy due to the intermediate or complex habitats in the field.

The Arctic Vegetation Map

Bioclimatic map is different from the vegetation map. Bioclimatic subzones represent a framework of climate effecting on floras and specific vegetation types. Vegetation is affected by not only climate but also other environmental factors such as soil properties, topography, altitude, the direction of slope, drainage, and so on. Therefore, a bioclimatic zone can include different vegetation types.

The Arctic vegetation map adopted 15 vegetation units based on plant functional types (Table 2.4). Plant growth form (e.g., graminoids, shrubs), size (e.g., dwarf and low shrubs), and taxonomical status (e.g., sedges, rushes, grasses) were the basis for

Table 2.4 Area of the Arctic vegetation units

Vegetation unit	Area ($\times 1000$ km^2)	%
Barrens	1358	26.79
• B1. Cryptogam, herb barren	225	4.44
• B2. Cryptogam complex (bedrock)	372	7.34
• B3. Non-carbonate mountain complex	625	12.33
• B4. Carbonate mountain complex	136	2.68
Graminoid tundras	1475	29.09
• G1. Rush/grass, forb, cryptogam tundra	141	2.78
• G2. Graminoid, prostrate dwarf-shrub, forb tundra	429	8.46
• G3. Non-tussock sedge, dwarf-shrub, moss tundra	569	11.22
• G4. Tussock-sedge, dwarf-shrub, moss tundra	336	6.63
Prostrate-shrub tundras	539	10.63
• P1. Prostrate dwarf-shrub, herb tundra	399	7.87
• P2. Prostrate/hemi-prostrate dwarf-shrub tundra	140	2.76
Erect-shrub tundras	1302	25.68
• S1. Erect dwarf-shrub tundra	689	13.59
• S2. Low-shrub tundra	613	12.09
Wetlands	396	7.81
• W1. Sedge/grass, moss wetland	101	1.99
• W2. Sedge, moss, dwarf-shrub wetland	136	2.68
• W3. Sedge, moss, low-shrub wetland	159	3.14
Total	5070	100

This area does not include the area covered by glaciers (http://www.arcticatlas.org/maps/themes/cp/cpvg)

distinguishing the vegetation units. The functional types of dominant plants, floristic province, landscape types, lake cover, and substrate chemistry were considered to make the 15 units. The dominant bedrock (e.g., carbonate and noncarbonate) was also used for mountains, because it has complexes of vegetation.

The legend of 15 vegetation units contains five categories: B = barrens; G = graminoid-dominated tundras; P = prostrate-shrub-dominated tundras; S = erect-shrub-dominated tundras; W = wetlands. The unit names consist of the alphabetic codes with numeric codes. Graminoid tundras cover the largest area (29.1% of the map), followed by barrens (26.8%), erect-shrub tundras (25.7%), prostrate-shrub tundras (10.6%), and wetlands (7.8%).

Barrens

- B1. Cryptogam, herb barren
- B2. Cryptogam complex (bedrock)
- B3. Noncarbonate mountain complex
- B4. Carbonate mountain complex

Graminoid Tundras

- G1. Rush/grass, forb, cryptogam tundra
- G2. Graminoid, prostrate dwarf-shrub, forb tundra
- G3. Non-tussock sedge, dwarf-shrub, moss tundra
- G4. Tussock-sedge, dwarf-shrub, moss tundra

Prostrate-Shrub Tundras

- P1. Prostrate dwarf-shrub, herb tundra
- P2. Prostrate/hemi-prostrate dwarf-shrub tundra

Erect-Shrub Tundras

- S1. Erect dwarf-shrub tundra
- S2. Low-shrub tundra

Wetlands

- W1. Sedge/grass, moss wetland
- W2. Sedge, moss, dwarf-shrub wetland
- W3. Sedge, moss, low-shrub wetland

Relationship Between Arctic Vegetation Units and Bioclimatic Subzones

The first vegetation map covering the Arctic was published in 2003 (Fig. 2.11). This map used a single and unified legend to explain the Arctic vetation type. Bioclimatic subzones and vegetation units do not exactly coincide. Even under the similar climate (temperature and precipitation), vegetation can be different, because the plants also depend on the characteristics of the soil, drainage, wind intensity, slope, and the direction of the land are different.

Subzone A is predominantly glacier, including glacier areas (48% of the subzone), followed by cryptogam, forb barrens (B1, 24%). There are also rush/grass cryptogam tundra (G1, 19%) and non-carbonate mountain complexes (B3, 7%).

Subzone B also has large barren areas: cryptogam, forb barren (B1, 24%), mountain complexes (B3 and B4, 23%), and glaciers (12%). Vegetated areas have a mixture of rush/grass cryptogam tundra (G1, 17%), prostrate dwarf-shrub, herb tundra (P1, 11%), and graminoid, prostrate dwarf-shrub, forb tundra (G2, 9%).

In Subzone C, barren areas reduced less 40%: cryptogam, barren (B1 and B2, 18%), mountain complexes (B3 and B4, 12%), and glaciers (10%). Graminoid, prostrate dwarf-shrub, and forb tundra (G2, 23%) is most abundant, followed by

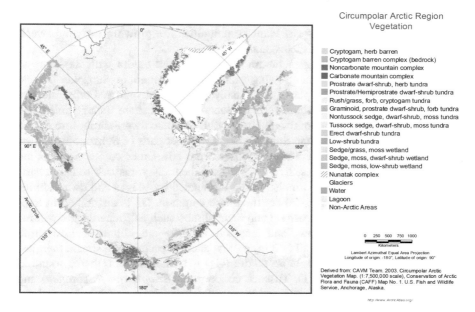

Fig. 2.11 Circimpolar Arctic vegetation map (CAVM Team (2003), © The Alaska Geobotany Center)

prostrate dwarf-shrub, herb tundra (P1, 16%), prostrate/hemi-prostrate dwarf-shrub tundra (P2, 8%), and wetlands (W1, 6%).

Subzone D is much more vegetated than Subzone C, with large areas of non-tussock sedge, dwarf-shrub, moss tundra (G3, 25%). Erect shrub tundras (S1 and S2) cover 20%, wetlands (W2 and lakes) 11.5%, and prostrate-shrub tundras (P1 and P2), 10%. Mountain complexes (B3 and B4) cover 12% and barren shield areas (B2) 10%.

Subzone E is the most densely vegetated subzone with over half of the subzone covered by erect shrub vegetation (S1 23%, and S2 30%). Tussock-sedge, dwarf shrub, moss tundra (G4) covers 13%, wetlands (W3 and lakes) 11.3%, and mountain complexes (B3 and B4) 11%.

We can find "cryptogam, herb barren (B1) unit" in Subzones A and B, some C at higher elevations. Herbs, mosses, liverworts, and lichens are scattered sparsely, and the most common vascular plants are cushion forbs and graminoids in B1. Cryptogam barren complex (B2) is found in Subzones C and D. Rocks covered by lichens are interspersed with lakes and more vegetated areas. Alpine sweetgrass, dwarf shrubs, and fruticose lichens grow between the bedrock outcrops. Non-carbonate mountain complex (B3) is everywhere in the Arctic Subzones A through E, and carbonate mountain complexes (B4) is distributed in Subzones B through E with limestone or dolomite bedrock. Plant communities are growing on wind-swept, rocky ridges, screes, dry fell-fields, and snowbed.

Rush/grass, forb, cryptogam tundra (G1) is a moist tundra with moderate to complete cover of very low-growing plants. G1 is found in Subzones A and B. The dominant vascular plants are usually grasses and rushes. Forbs, mosses, liverworts, and lichens are common. In Subzone B, prostrate dwarf shrubs and sedges are present but not dominant. Graminoid, prostrate dwarf-shrub, forb tundra (G2) is a moist to dry tundra in Subzone C and warmer parts of Subzone B. Sedges are dominant, along with prostrate shrubs <5 cm tall. Other common plants are rushes grasses, forbs, mosses, liverworts, and lichens. Non-tussock sedge, dwarf-shrub, moss tundra (G3) is a moist tundra dominated by sedges and dwarf shrubs <40 cm tall. Moss layers (5~20 cm thick) are common. G3 is distributed in Sub-zones C through E. Hemi-prostrate and erect shrubs such as Richardson's willow, gray willow, tealeaf willow, and Lapland rosebay are common. Low-shrub (40~200 cm tall) and some tall (>2 m) willow thickets occur along stream margins. Tussock-sedge, dwarf-shrub, moss tundra (G4) is a moist tundra, that is found in Subzone E, some Subzone D. Tussock cottongrass (*Eriophorum vaginatum*) and dwarf shrubs <40 cm tall are dominated. Tussock sedges with other sedges, prostrate and erect dwarf-shrubs, and mosses are abundant in G4. Other common plants include grasses, forbs, and lichens.

Prostrate dwarf-shrub, herb tundra (P1) is a dry tundra with prostrate shrubs <5 cm tall (such as mountain avens and Arctic willow), graminoids, forbs, and lichens. P1 is distributed in Subzones B and large dry areas in C. Prostrate dwarf-shrubs are dominant. Prostrate/hemi-prostrate dwarf-shrub tundra (P2) is a moist to dry tundra in Subzone C. Prostrate and hemi-prostrate shrubs <15 cm tall, particu-larly mountain heather are dominant. Rushes, forbs, mosses, and lichens are also common.

Erect dwarf-shrub tundra (S1) is a moist to dry tundra in Subzone D on acidic soils. Hemi-prostrate and erect dwarf shrubs <40 cm tall are dominant. Drier, lichen-rich dwarf-shrub tundras are common in many areas. Low-shrub tundra (S2) is a moist to dry tundra in Subzone E. Low shrubs >40 cm tall are dominant. Thick moss carpets are common in most shrublands. Low and tall willows and alders are abundant along drainages and near tree line. Some trees reach into this subzone along the southern river valleys. Peatlands with permafrost are common in wet areas.

Sedge/grass, moss wetland (W1) is wetland complexes of Subzones B and C, including water, low wet areas, and moist elevated microsites. Sedges, grasses, and mosses are dominant. Sedge, moss, and dwarf-shrub wetland (W2) are wetland complexes of Subzone D. Sedges, grasses, and mosses are dominant, and prostrate dwarf-shrubs and forbs are often present. Sedge, moss, and low-shrub wetland (W3) are wetlands in Subzone E. Sedges, moss, and low shrubs >40 cm tall are dominant. The main plant communities on elevated microsites are shrublands with prostrate and erect dwarf-shrubs and mosses (Fig. 2.12).

Fig. 2.12 Front vegetated tundra and rear barren tundra

Bibliography

CAFF (Conservation of Arctic Flora and Fauna) (2010) Arctic Biodiversity Trends 2010 – Selected indicators of change

CAVM (Circumpolar Arctic Vegetation Map) Team (2003) Circumpolar Arctic Vegetation Map (1:7,500,000 scale), Conservation of Arctic Flora and Fauna (CAFF) Map No. 1. U.S. Fish and Wildlife Service, Anchorage, Alaska. http://www.geobotany.uaf.edu/cavm/

Chernov YI, Matveyeva NV (1997) Arctic ecosystems in Russia. In: Wielgolaski F-E (ed) Polar and alpine tundra. Ecosystems of the world 3:361–508

Daniëls FJ, Bültmann H, Lünterbusch C, Wilhelm M (2000) Vegetation zones and biodiversity of the North-American Arctic. Berichte der Reinhold-Tüxen-Gesellschaft 12:131–151

Edlund S (1990) Bioclimate zones in the Canadian Archipelago. In: Harrington CR (ed) Canada's missing dimension: science and history in the Canadian Arctic Islands. Canadian Museum of Nature, Ottawa, pp 421–441

Elvebakk A (1985) Higher phytosociological syntaxa on Svalbard and their use in subdivision of the Arctic. Nord J Bot 5:273–284

Elvebakk A (1999) Bioclimatic delimitation and subdivision of the Arctic. In: Nordal I, Razzhivin VY (eds) The species concept in the High North – A Panarctic Flora initiative. The Norwegian Academy of Science and Letters, Oslo, pp 81–112

Elvebakk A, Elven R, Razzhivin VY (1999) Delimitation, zonal and sectorial subdivision of the Arctic for the Panarctic Flora Project. In: Nordal I, Razzhivin VY (eds) The species concept in the High North – A Panarctic Flora initiative. The Norwegian Academy of Science and Letters, Oslo, pp 375–386

Nuttall M (ed) (2005) Encyclopedia of the Arctic. Routledge, London, pp 117–121

Polunin N (1951) The real Arctic: suggestions for its delimitation, subdivision and characterization. J Ecol 39:308–315

Raynolds MK, Comiso JC, Walker DA, Verbyla D (2008) Relationship between satellite-derived land surface temperatures, arctic vegetation types, and NDVI. Remote Sens Environ 112:1884–1894

Sale R (2008) The Arctic: the complete story. Frances Lincoln, London, pp 15–21

Tuhkanen S (1986) Delimitation of climatic-phytogeographical regions at the high-latitude area. Nordia 20:105–112

Walker DA (2000) Hierarchical subdivision of arctic tundra based on vegetation response to climate, parent material, and topography. Glob Chang Biol 6(S1):19–34

Walker DA, Gould WA, Maier HA, Raynolds MK (2002) The circumpolar Arctic vegetation map: AVHRR-derived base maps, environmental controls, and integrated mapping procedures. Int J Remote Sens 23:4551–4570

Young SB (1971) The vascular flora of St. Lawrence Island with special reference to floristic zonation in the arctic regions. Contrib Gray Herb 201:11–115

Yurtsev BA (1994) The floristic division of the Arctic. J Veg Sci 5:765–776

Chapter 3
Arctic Plants in Different Tundras

When I visited Svalbard for the first time in 2003, the plants were so tiny that I could barely see them. Arctic tundra seemed to be an environment too harsh for plant growth. Seven years later, I was surprised to see the field full of various plants on Seward Peninsula in Alaska. I wondered how these many plants lived in the Arctic.

There are 300,000 species of plants on Earth, and there are around 3000 species of plants living in the Arctic, including seed plants and mosses. The number of Arctic seed plants is 1600~2000 including subspecies. A total of 1300 mosses and liverworts live in the Arctic. In numbers, about 1% of plants in the world live in the Arctic. Compare to more temperate ecosystems, the diversity of Arctic vegetation is very low. Nonetheless, they support large populations of arctic animals such as reindeer, muskox, and seasonal birds by supplying foods and nesting habitats.

Most Arctic plants grow close to the ground. In so doing, they secure a blanket of snow to envelop them and protect them in cold air. Some of them grow densely forming mats or cushions, and others produce hairs covering leaves and stems. Most Arctic vascular plants are specific to tundra, but some of them can live in taiga and High Mountain of temperate region. Arctic plants find their own ways to overcome the harsh environment.

We already looked at the life forms of Arctic plants in Table 2.1. However, when we read a book or papers on Arctic plants, we find other terms explaining Arctic vegetation: Desert tundra, dry tundra, mesic tundra, moist tundra, wet tundra, heath tundra, tussock tundra, meadow tundra, steppe tundra, shrub tundra, and so on. The preceding five terms are related to the moisture content of the soil. Dry tundra, moist tundra, and wet tundra often include the following kinds of plants: dry heath tundra, moist tussock tundra, and wet sedge tundra. Let us take a closer look at Arctic plants according to the different tundra regions: polar desert, dry tundra, mesic tundra, moist tundra, wet tundra, and shrub tundra.

© Springer Nature Switzerland AG 2020

Y. K. Lee, *Arctic Plants of Svalbard*, https://doi.org/10.1007/978-3-030-34560-0_3

The Persistent Plants in Polar Desert

Desert tundra is better known as polar desert. Polar desert in the Arctic is the most northern region where it is sparsely vegetated. Polar desert is an arid land that has less than 250 mm of annual rainfall. Polar semidesert has often more precipitation or less drainage than polar desert. Both polar desert and semidesert belong to the High Arctic. The land surface of polar desert consists of rocks and rubbles, exposed bedrock, and bare mineral soil (Fig. 3.1). Short growing season, low temperatures, lack of water, infertile soils, and harsh wind inhibit plant growth in polar desert.

Vascular plants forms scattered patches, and the plant coverage is usually less than 5%. In polar semidesert, vascular plants contribute 5~20% cover and mosses and lichens contribute 5~80%. Polar desert plants in the Arctic are short, and the height of the plants is generally less than 2 cm. We can find the same species or genus of many tundra plants elsewhere in similar Arctic environments. About 90% of the vascular plants in polar desert are circumpolar.

Flowering plants such as Arctic poppy, saxifrage, moss campion, and bluegrasses are dominant in the polar desert (Fig. 3.2). Nonflowering liverworts and mosses, and spore forming photosynthetic organisms such as lichens and algae are also dominant in polar desert. We can also observe mat-forming plants such as mountain avens, dwarf willow, draba, and Arctic bell-heather here.

In the polar desert, some places are exceptionally full of plants like the oases. These places have a well-drained soil where plants can take root, a good topography that is protected from the wind, or fertile soil where animal bodies and feces are. It shows greater productivity, and higher species diversity and/or abundance.

We can see a thin film like mocha crust covering the surface of the soil in polar desert. It is "biological soil crust," and often called "black crust" because it is mainly

Fig. 3.1 Bedrock exposed area in Svalbard

Fig. 3.2 Tufted saxifrage in a Svalbard polar desert

black (Fig. 3.3). The biological soil crust is a small living ecosystem with a population of cyanobacteria, fungi, lichens, and moss. Biological soil crust can survive in extreme environments because they can withstand long periods of drying, extreme temperatures, intense light, and so on. The formation of a thin membrane is due to the accumulation of soil particles with filamentous cyanobacteria in the polysaccharide secreted by microorganisms. Because the soil is fixed in this way, biological soil crust helps stabilize the soil and prevent soil erosion. Cyanobacteria and lichens in biological soil crust can fix nitrogen in the atmosphere. In addition, the bottom of the biological soil crust can be a habitat for various organisms by supplying nutrients to the soil and increasing the moisture retention in the soil. This plays an important role in the barren ecosystem.

Dry Heath Tundra

Dry tundra has well-drained dry soils with thin organic mats (<5 cm), deep thaw depths, and heath plants along with lichens, dwarf shrubs, or cushion plants. Dry tundra is a wide range of tundra habitats such as ridge tops, stony soil in rocky habitats, or steeply sloping land. Dry tundra are often exposed to strong wind and desiccation, and limited snow cover during winter. The plant species are similar to those of the polar desert, but sometimes with a more extensive coverage. Due to the thin snow cover in dry tundra, animals such as reindeer, muskox, and snow sheep can find food from the dry tundra pasture in winter.

The common plant species found in dry tundra are oxytrope, Arctic cinquefoil, alpine bearberry, lingonberry, and Lapland diapensia. Prostrate plants like mountain

Fig. 3.3   Biological soil crust in a Svalbard polar desert

Fig. 3.4   Prostrate and dwarf shrubs on a rocky mountain in Troms, Norway

avens, alpine azalea, alpine bearberry, and some willows are common here (Fig. 3.4).
Low dwarf shrubs of bilberry, cowberry, Labrador Tea, and Arctic bell-heather are
frequent. The berry-producing heath plants such as Northern bilberry, rock
branberry, and cowberry increase in the locations with limited snowfall. Arctic
bell-heather dominates with an admixture of prostrate willow in well-drained
snow-rich habitats.

Many dry tundra plants are covered by downy and short hair, which helps retain
water and heat emitted by the plant (Fig. 3.5). Some plants form compact mat of

Fig. 3.5 Arctic willow covered by hairs which is living in Zackenberg, Greenland

cushion that helps them keep warm several degrees above the air temperature. Arctic cushion plants usually have small and fleshy leaves that reduce the surface area of the plant, which reduces transpiration of water.

Heath is a common vegetation in dry tundra; therefore, dry tundra is often called as dry heath tundra. Representative heaths are *Cassiope* heath and *Empetrum* heath. *Cassiope* heath is a circumpolar vegetation type predominated by Arctic bell-heather (Fig. 3.6). The scientific name of Arctic bell-heather is *Cassiope tetragona*, of which the common name is Arctic white heather or white Arctic mountain heather. The name of "Arctic bell-heather" came from the bell-shaped white flowers. Arctic bell-heather is an evergreen dwarf shrub with dark green leaves. The leaves are arranged in four rows and cover the plants. Although the leaves of Arctic bell-heather die off after 3~4 years, and they remain attached on the plant for several decades. The number of leaves and flowers produced each year positively correlated with the July mean temperature. Therefore, Arctic bell-heather is potentially useful as a bioindicator in climate change research.

Empetrum heath is a circumpolar vegetation dominated by species of the genus *Empetrum*. The name of *Empetrum* came from the Greek *en petros*, meaning "on rock." Crowberry is a representative plant in Empetrum heath (Fig. 3.7). The scientific name of crowberry is *E. nigrum*, which has two subspecies: *E. nigrum* ssp. *hermaphroditum* and *E. nigrum* ssp. *nigrum*. Both have long, creeping branches with needlelike leaves. The length of leaves is different; long leaves of *E. nigrum* ssp. *hermaphroditum* are up to 3 times as long as they are wide, but those of *E. nigrum* ssp. *nigrum* are 3~5 times as long as wide. The black berries of crowberry persist on the branches over winter, which are edible by mammals and birds.

Fig. 3.6 The white flowers and red fruits of Arctic bell-heather

Crowberry leaves have chemical substances, which restrain germination and growth of other plants. They also contain toxins that inhibit reindeer and other herbivores from eating them.

Fig. 3.7 Crowberry with black and juicy fruits

Mesic Tundra and Moist Tussock Tundra

Mesic tundra is an intermediate habitat between dry tundra and wet tundra. Mesic tundra is usually covered by snow during winter. Mesic tundra maintains moist conditions due to high ice content of permafrost that inhibit draining water. We can find various shrubs, sedges, tussock, mosses, and lichens in mesic tundra. The thickness of the active layer correlates with the dominant vegetation. Tussock is abundant in a flat terrain with 40~60 cm deep of active layer. The active layer depth is 60~90 cm for dwarf shrub, and ~130 cm for low shrub.

The surface of mesic tundra is often covered by moss and liverwort such as bog groove-moss, turgid Aulacomnium moss, elongate Dicranum moss, Girgensohn's bog-moss, glittering wood-moss, Juniper haircap, red-stemmed feather-moss, woolly feather-moss, and ciliated fringewort. Mesic tundra is a habitat for various lichens such as reindeer lichen, green reindeer lichen, Arctic finger lichen, Green Witch's hair, white worm lichen, *Flavocetraria nivalis*, and *F. cucullata*. Lichen is also an important food for reindeer.

The dwarf birch, a representative dwarf shrub in mesic tundra, is with woolly willow and downy willow. We can find mountain alder and Siberian alder on southern mesic tundra. Berry-producing plants such as Arctic Raspberry, cloudberry, crowberry, lingonberry, and alpine bearberry grow in mesic tundra. They are important foods for reindeer and snow sheep. Dwarf shrubs such as Lapland Diapensia, Lapland rosebay, and Labrador tea, and prostrate willows such as arctic willow and gray willow are also abundant in mesic tundra.

Moist tundra occupies gently sloping land with dense organic mats that thaws into the mineral soil during most summers. The depth of active layer is about 40–60 cm, and the thickness of the active layer is the major environment constraint of

Fig. 3.8 Dry dead leaves make base of a tussock

cottongrass tussock tundra. Tussock-forming sedges dominated with admixtures of dwarf shrubs, mosses, lichens, and forbs in moist tundra. Therefore, it is often called as moist tussock tundra. The name is derived from the dominance of the cottongrass tussock, *Eriophorum vaginatum*. Tussock tundra also includes other tussock-forming sedge and dwarf shrubs, lichen, and mosses.

Tussock tundra has a circumpolar distribution, but is most frequent in areas where the active layer of the permafrost is about 50 cm deep. Tussocks are formed when the dead leaves of cotton grass and sedges take time to decompose in the cool and acidic waterlogged ground at the base of the plants (Fig. 3.8). Dead material therefore builds up at the plant base. Eventually this material is converted to soil, and as more is added, it breaks the water surface. It is then exploited by other plants, dead leaves from these being added to the base so the tussock height increases. Individual tussock wobbles and is unstable, making tussock hopping a risky means of travel. The ground between the tussocks is waterlogged, the water often overtopping a walker's boot. The tussock tundra is the ideal breading place for mosquitoes. Tussocks also burn dead, dry leaves as its tussock base make good tinder. As new buds of cotton grasses and other plants are often buried deep inside the tussock to survive, the flames become a useful regenerator.

Dwarf birch, tealeaf willow, bog bilberry, mountain cranberry, marsh Labrador tea, and black crowberry are common in tussock tundra. The most common mosses include glittering wood moss, elongate Dicranum moss, turgid aulacomnium moss, and woolly feather moss. Sphagnum mosses are also abundant, especially where drainage is poor. The lichens *Flavocetraria nivalis, F. cucullata, Cladonia rangiferina, Dactylina arctica*, and *Thamnolia vermicularis* are also common. Tussock tundra is an important summer range for reindeers.

Wet Sedge Tundra

Wet tundra covers around half of northern Siberia, large area of the central Canadian mainland, and much of northern Alaska. It is also predominant in many places of the Low Arctic. Wet tundra is poorly drained lowlands with huge peat accumulations (>40 cm, often below permafrost). Wet tundra is dominated by sedges and mosses, and is often called "wet sedge tundra." The dominant plants are tall cottongrass, hare's tail cottongrass, white cottongrass, water sedge, and looseflower alpine sedge (Fig. 3.9).

Wet sedge tundra is less uniform than tussock tundra since it is less dependent on a single species. Cottongrass is typically abundant or even dominant. The most common cottongrass of wet tundra on a circumpolar basis is tall cottongrass, and the most abundant sedge is water sedge. Tall cottongrass, white cottongrass, and Fisher's tundra grass are also common. Wet lands are important for water storage in the water-poor environment after the first few weeks of spring melt.

Most vascular plants are perennial, long-lived species ranging in age from 5 to 7 years for individual graminoid stems to 5~100+ years for tussock graminoid and deciduous shrubs. Surface hydrology is extremely important for the development and maintenance of tundra.

Fig. 3.9 Tall cottongrass

Shrubs Near Tree Line

Close to the tree line the shrubs, particularly the birch, willow, and alder species, grow taller and further species—aspens and poplars—become established, creating an area of forest tundra. Interestingly, forest tundra often has fewer species than either the tundra to the north or the boreal forest to the south. In forest tundra, berry-producing shrubs tend to grow taller and set more fruit, making the area particularly attractive to reindeer. Rushes are found in the wetter areas. The tree line is the transition between boreal forest and tundra ecosystems. It is a major global biogeographic boundary, separating the circumpolar boreal forest (Subarctic) from the Arctic tundra. In North America, black spruce and white spruce form the tree line. Spruce, larch, pine, and birch all form tree line in different regions.

Shrub tundra communities are usually well stratified vertically into three layers. The shrub layer is dominated by dwarf birch, and several willow shrubs (tea-leaved willow, downy willow, woolly willow, thin red willow). In the southernmost part of tundra where it is open woodland, the alder shrub of *Alnus fruticose* is common. The middle layer consists of abundant dwarf-shrubs such as crowberry, mountain cranberry, black bearberry, Labrador tea, and herbs such as cottongrass, cloudberry, and several sedges forming a continuous herb–dwarf–shrub layer. The ground layer consists of moss (Bog groove-moss, Elongate Dicranum moss, Girgensohn's bog-moss, Glittering wood-moss, Red-stemmed feather-moss, and Turgid Aulacomnium moss) and lichens (Reindeer lichen, Green reindeer lichen, *Flavocetraria nivalis*, and *F. cucullata*) (Fig. 3.10). Shrub tundra is a major reindeer pasture.

Toward the south, tree line resembles the boreal forest, whereas toward the north, it resembles the tundra (Fig. 3.11). The geographic location of the tree line varies

Fig. 3.10 Lichen known as reindeer moss

Fig. 3.11 The border of tree line in Seward Peninsula, Alaska

through time: it can move northward when climate conditions are favorable, and retreat southward under less favorable conditions. The processes involving forthward and southward movement are, however, quite different, so that the nature of advancing (northward movement) and retreating (southward movement) tree lines are quite different. The position of the tree line moves northward when the climatic conditions in the tundra become favorable for tree growth. The conditions permitting establishment of a tree vary among species, but in general, trees need a certain length of growing season, sufficient precipitation, and warm enough conditions during the growing season. Winter temperatures are less important, provided the tree was able to harden in the autumn or late summer. Snow depth is important, and sufficient snow is needed to protect the needles from desiccation or abrasion. Frequently, at the northern limit of a tree species, trees only grow in more protected areas when there is an accumulation of snow. Southward movements of the tree line occur by very different dynamics. If the environment should be unfavorable for tree reproduction in the forest tundra, the trees can probably persist for a long-term period, although growth may be reduced and there may be no reproduction by seeds. The trees may adopt a shrub form and reproduce by layering.

Bibliography

Consortium of North American Lichen Herbaria. http://lichenportal.org/portal/

Elven R, Murray DF, Razzhivin VY, Yurtsev BA (2007) Checklist of the Panarctic flora (PAF). Vascular plants. University of Oslo, Norway. http://nhm2.uio.no/paf/

Nuttall M (ed) (2005) Encyclopedia of the Arctic. Routledge, London, pp 513–529, 1279–1280, 2067–2072, 2169–2171

Sale R (2008) The Arctic: the complete story. Frances Lincoln, London, pp 273–284

Chapter 4
Arctic Plants in Svalbard

Svalbard, Cold Shore

Svalbard is not a name of an island. Svalbard is the name of an archipelago that includes all of the islands and islets of the Arctic Ocean. The range of Svalbard archipelago extends between 74°~81° north latitude, and 10°~35° east longitude. If you want to find Svalbard on the world map, you may find a bunch of islands from the top right of Greenland.

Svalbard, meaning "cold shore" or "cold edge" in ancient Icelandic, is a very old name. They found "Svalbarði fundinn (Svalbard was discovered)" in Icelandic annals published over 800 years ago. But it is not clear that which island in Svalbard this name was indicating. It may not be one of Svalbard archipelago, because Svalbard had also been used to refer to the uninhabited parts of Greenland.

Spitsbergen is the main island of Svalbard. It was officially discovered by Willem Barentsz on his third expedition to find the Northeast Passage to China in June 1596. Barentsz wrote, "The terrain was mostly broken up, fairly elevated, and consisted of mountains and craggy peaks; therefore, we named it Spitsbergen." Until the 1920s, Spitsbergen referred to the archipelago as well as one of its islands. Thus, there was some confusion between Svalbard and Spitsbergen. In 1925, Norway distinguished between the name of archipelago (Svalbard) and the main island (Spitsbergen). There are many islands in Svalbard. Spitsbergen is the largest one, and the second is Nordaustlandet, meaning "North East Land," which is located northeast of Spitsbergen (Fig. 4.1). The third one is Edgeøya, meaning "Edge Island," which lies southeast of Spitsbergen. On the southeast of Edgeøya, a small needle-shaped island, Hopen inclines. The southernmost island of Svalbard is Bjørnøya, meaning "Bear Island." This small island looks like water drop upside down, and lies between Spitsbergen and Norway mainland. People live on Spitsbergen only, and few staff visit regularly Bjørnøya and Hopen to maintain weather stations.

Svalbard is very dry, and the annual precipitation is only 400 mm on the west coast. There are many arid regions in the Arctic, which are called the Arctic Desert or

© Springer Nature Switzerland AG 2020
Y. K. Lee, *Arctic Plants of Svalbard*, https://doi.org/10.1007/978-3-030-34560-0_4

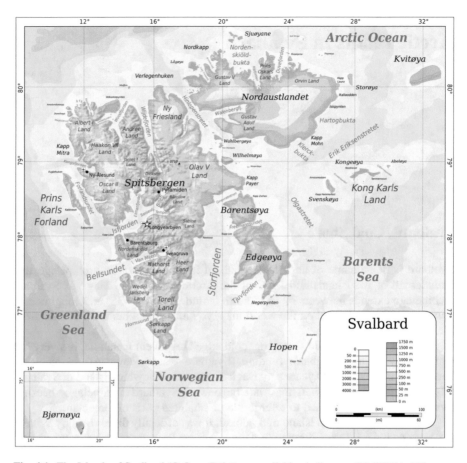

Fig. 4.1   The Islands of Svalbard (© Oona Räisänen, available via license: CC BY-SA 4.0)

the Tundra Desert. In the summer, we stay in Longyearbyen or Ny-Alesund from 2 weeks to 2 months in short, in order to conduct research. When I went to bed at night, my body could feel how dry this place was. So I put a humidifier or hang a wet towel in the room. I used to apply lip balm and plenty of moisture cream, which I do not use usually. It is not only people who lack moisture from the dry atmosphere, plants also must cope in this dry environment.

Thanks to the warm North Atlantic current, the climate is not so harsh compared to the inland Arctic. On the west coast (Isfjord radio), the mean temperature for the winter (from January to March) over the last 30 years was around −12 °C. Summers are also mild: the mean temperature for July was around 5 °C on the west coast on Svalbard.

The first plant sampling on Svalbard was conducted about 200 years ago. Baltazar Keilhau, a Norwegian geologist explored Edgeøya, the third largest island in

Svalbard archipelago, and brought a bundle of plant samples in 1827. Some of the plant samples were investigated and reported by Sommerfelt in 1832. After that, many botanist have visited many places of Svalbard and collected plant samples to make herbarium. Among them, Hanna Marie Resvoll-Holmsen was the first woman botanist who joined the 1907 expedition supported and led by the oceanographer Albert I, Prince of Monaco. As a student at University of Oslo, Hanna made vegetation maps for plants living on the western Spitsbergen. She also discovered several new species there. This expedition pushed the Norwegian government to support their scientific cruises in 1908, and Hanna Marie Resvoll-Holmsen participated in the following expedition too. She took many pictures—color photographs—of Svalbard during her second visit. She published a wonderful book titled "Svalbard's flora," which was the first comprehensive book to cover the flora of Svalbard. Since then, Svalbard's flora has been continuously reported on, and the masterpiece was published by Rønning, and Elven and Elvebakk in 1996, respectively. In a digital era, a fantastic Web (http://svalbardflora.no/) provides specific information on vascular plants in Svalbard. According to the Flora of Svalbard Web, 184 species are living on Svalbard. When I checked and compiled the lists of Svalbard flora, more than 200 vascular plants have been reported, and the list of the plants is in the appendix. Vascular plants do not include mosses and lichen.

Among the Svalbard plants, 48 are rare species listed on the Red List (Table 4.1). Elven and Elvebakk (1996) also checked the rare and endemic (or highly disjunct) species in Svalbard: 49 plants are very rare and 35 plants are living on Svalbard only. After several decades, we may no longer be able to see these plants.

When I visited Svalbard for the first time, I could not identify the plants in brownish ground. They appeared too tiny for me to distinguish from the sandy soil surface. At the time, I was not able to catch any of the surroundings because I was walking too fast. But as I slowed down and took my time to look carefully at the unfamiliar environment, I could begin to spot all sorts of flowers.

When people arrive at the Longyearbyen airport, they usually wait for a bus at the gate of the airport to go to the hotel. But I always choose to walk down along the seashore from the airport so that I can enjoy seeing the summer flowers in yellow, white, and pink. They are amazing pieces of wonder that bloom—overcoming the harsh coldness and wind with so little nutrient.

Plants Without Flowers

Catherine La Farge is a curator of Cryptogamic Herbarium at the University of Alberta in Canada. Cryptogam is an old word that originates from the Greek word, "kryptos gameein," meaning hidden reproduction, which referred to plants without seed. Linné, father of modern taxonomy, had divided the plant kingdom into 24 classes, and one of them was the "Cryptogamia." Cryptogamia included four different groups: algae, bryophytes, ferns, and fungi in Linne's age. Algae and fungi no longer belong to plants. Cryptogam does not bloom flowers and has no seeds. It

Table 4.1 Svalbard vascular plants on the Red List

Scientific name	Red List	Rare, endemic; Synonym (Reference)
Lycopodiophyta Ophioglossales Ophioglossaceae		
Botrychium boreale	CR	Rare, endemic
Botrychium lunaria	CR	Rare, endemic
Polypodiales Woodsiaceae		
Woodsia glabella	EN	Rare
Magnoliophyta Poales Cyperaceae		
Carex aquatilis var. *minor*	CR	Rare; *Carex concolor* (The Flora of Svalbard); *C. stans* (Rønning 1996); *Carex aquatilis* ssp. *stans* (Elven and Elvebakk 1996)
Carex bigelowii ssp. *arctisibirica*	CR	
Carex capillaris ssp. *fuscidula*	VU	Rare, endemic; *Carex capillaris* (Elven and Elvebakk 1996; Rønning 1996)
Carex glacialis	VU	Rare
Carex krausei	VU	Rare, endemic
Carex lidii	VU	
Carex marina ssp. *pseudolagopina*	VU	Rare, endemic
Eriophorum × *rousseauianum*	NT	*Eriophorum sorensenii* (The Flora of Svalbard)
Kobresia simpliciuscula ssp. *subholarctica*	EN	Rare, endemic; *Kobresia simpliciuscula* (Elven and Elvebakk 1996; Rønning 1996)
Poales Juncaceae		
Juncus arcticus	VU	Rare
Juncus castaneus ssp. *leucochlamys*	EN	*Juncus leucochlamys* (The Flora of Svalbard)
Luzula wahlenbergii	NT	
Poales Poaceae		
Arctagrostis latifolia	EN	
Arctodupontia scleroclada	EN	
Calamagrostis purpurascens	VU	
Festuca brachyphylla	VU	Rare, endemic
Festuca hyperborea	NT	
Pleuropogon sabinei	NT	
Puccinellia distans	NT	*Puccinellia coarctata* (The Flora of Svalbard, Rønning 1996)

(continued)

Table 4.1 (continued)

Scientific name	Red List	Rare, endemic; Synonym (Reference)
Puccinellia tenella	CR	Rare, endemic; *Puccinellia svalbardensis* (The Flora of Svalbard, Elven and Elvebakk 1996; Rønning 1996)
Puccinellia vahliana	NT	*Colpodium vahlianum* (Dahl 1937; Neilson 1968, 1970; Rønning 1996)
Pucciphippsia vacillans	NT	Endemic; *Colpodium vaeillans* (Dahl 1937; Neilson 1968, 1970; Rønning 1996)
Ranunculales Ranunculaceae		
Beckwithia glacialis	VU	Endemic; *Ranunculus glacialis* (Elven and Elvebakk 1996; Rønning 1996)
Coptidium pallasii	NT	*Ranunculus pallasii* (Rønning 1996)
Ranunculus wilanderi	EN	Rare, endemic
Caryophyllales Caryophyllaceae		
Arenaria humifusa	VU	Endemic
Honckenya peploides	NT	*Honckenya peploides* ssp. *diffusa* (The Flora of Svalbard, Elven and Elvebakk 1996)
Minuartia rossii	NT	
Minuartia stricta	CR	Rare
Sagina caespitosa	EN	Rare, endemic
Brassicales Brassicaceae		
Draba fladnizensis	VU	
Draba micropetala	NT	
Draba pauciflora	NT	
Malpighiales Salicaceae		
Salix lanata	CR	
Rosales Rosaceae		
Alchemilla glomerulans	CR	Rare
Rubus chamaemorus	CR	
Sibbaldia procumbens	EN	Rare, endemic
Fagales Betulaceae		
Betula nana	NT	*Betula nana* ssp. *tundrarum* (The Flora of Svalbard)
Ericales Ericaceae		
Harrimanella hypnoides	NT	
Vaccinium uliginosum ssp. *microphyllum*	EN	Rare, endemic; *Vaccinium gaultherioides* (Rønning 1996)
Gentianales Gentianaceae		
Comastoma tenellum	EN	*Mertensia maritima* ssp. *tenella* (The Flora of Svalbard)

(continued)

Table 4.1 (continued)

Scientific name	Red List	Rare, endemic; Synonym (Reference)
Lamiales Scrophulariaceae		
Euphrasia wettsteinii	EN	
Asterales Asteraceae		
Erigeron eriocephalus	VU	*Erigeron uniflorus* ssp. *eriocephalus* (Evju et al. 2010)
Asterales Campanulaceae		
Campanula rotundifolia	VU	Rare, endemic; *Campanula rotundifolia* ssp. *gieseckiana* (The Flora of Svalbard, Elven and Elvebakk 1996; Alsos et al. 2004)
Campanula uniflora	VU	

Scientific names were checked in The Flora of Svalbard (http://svalbardflora.no)
CR, critically endangered; EN, endangered; VU, vulnerable; NT, near threatened; Rare, very rare species; Endemic, endemic or highly disjunct species (Elven and Elvebakk 1996)

reproduces by spores instead. The Cryptogamic Herbarium has more than 200,000 specimens of mosses, liverworts, hornworts, lichens, and fungi.

In 2007, Catherine La Farge visited Teardrop Glacier on Ellesmere Island. It was not her first visit to the island. She had visited northern Ellesmere Island to analyze moss flora back in 1983. She was an expert in examining moss of the high Arctic. When she walked along the marginal region of the Teardrop Glacier, which had been uncovered by ice for 3 years, she found several mosses having greenish edges. The mosses had been buried under ice for a long time and had only come to be exposed to light and fresh air from recent ice melting and glacial retreat. Catherine La Farge and her fellows checked the age of mosses using radiocarbon dating, which showed that the mosses were more than 400 years old. The subglacial mosses had made new lateral branches from the senescent stems. They were alive!

Mosses have the ability to grow into a whole plant from a tiny fragment composed of small number of cells. Thus, they cultured the mosses for 6 months or 1 year, and found regeneration phenomena from four moss species: Turgid aulacomnium moss (*Aulacomnium turgidum*), Fine Distichium (*Distichium capillaceum*), Candlesnuffer moss (*Encalypta procera*), and Great hairy screw-moss (*Syntrichia ruralis*). They seemed to be imperishable.

By traditional definition, moss was classified as a tiny, nonvascular plant with no flowers. Moss is called as bryophyte. When scientists tried to reveal the origin of plants, comparison of plants' morphology led to a debate about whether bryophytes consisted of a single group or multiple groups. Phylogenetic studies using molecular makers also began to be conducted from the chaos of controversy. Some researchers insisted that the bryophyte was a single taxon on the basis of chloroplast protein-coding genes. On the contrary, other researchers put forth that Bryophyte composed of three groups of liverworts, mosses, and hornworts using nuclear-encoded ribosomal 18S rRNA (small-subunit rRNA) gene sequences. Brent D. Mishler, a native

Fig. 4.2 Tufted Saxifrage growing on the bryophyte carpet

southern Californian interested in the phylogeny of green plants, described this ongoing debate as follows:

In many analyses (including the combined molecular and morphological analysis) the three major lineages (i.e., liverworts, hornworts, and mosses) appear to be paraphyletic with respect to the tracheophytes, with an indication that the mosses alone may be the sister group of the tracheophytes; however, in other analyses the "bryophytes" are supported as a monophyletic group. (Mishler et al. 1994)

Bryophytes are important plants in Arctic tundra environment. They provide the basis for the Arctic tundra ecosystem. Arctic bryophytes can live on tundra soil with little nutrient and they become the pioneer plants in the barren place. Dying bryophytes provide a nutrient-rich habitat to other plants. In wet places, they look like green carpets, carpets sustaining the harsh tundra environment (Fig. 4.2). Svalbard plants such as alpine chickweed, mountain avens, polar willow, and saxifrage used to settle and grew on the bryophytes. The major moss species forming carpets are turgid aulacomnium moss, racomitrium moss (*Niphotrichum* species), and elongate dicranum moss (*Dicranum elongatum*), and carpets forming liverwort are ciliated fringewort (*Ptilidium ciliare*), Hatcher's barbilophozia (*Barbilophozia hatcheri*), monster pawwort (*Tetralophozia setiformis*), Wenzel's notchwort (*Lophozia wenzelii*), *Chandonanthus setiformis*, *Sphenolobus minutus*, and *Tritomaria quinquedentata*.

The bryophytes of Svalbard were compiled by Frisvoll and Elvebakk in 1996. They listed 288 mosses and 85 liverworts and estimated that more than 400 bryophytes are living in Svalbard. Among them, 21 mosses and 14 liverworts were found only once. One out of four or five bryophytes was known as rare species, which were observed less than three times. Unfortunately, the bryophytes of Svalbard seemed unlikely to survive.

Fig. 4.3 Mountain fir-moss (Courtesy of Youngsim Hwang)

The rarest mosses on Svalbard, namely those that had been found only 1–3 times then have been listed by Frisvoll and Blom (1993); the list includes (35 species) that were found once, 10 liverworts and 15 mosses (25 species) were found twice, and 6 liverworts and 11 mosses (17 species) found were thrice in all 77 rare species. A few more rare species have been added in this chapter.

Clubmoss is not a bryophyte, even though its name includes moss. Clubmoss is different from the bryophytes in that it has a vascular system. In the taxonomic hierarchy, clubmoss falls in the Division Lycopodiophyta. It is also called lycopods too. The leaf of lycopods has a single vein (vascular trace), but the leaves of ferns and flowering plants have more complex veins. Only two plant lycopods have been recorded in Svalbard: Mountain fir-moss (*Huperzia arctica*) and mountain club-moss (*Lycopodium selago*). They are taller and wider than bryophytes. Their stems are covered by small leaves (Fig. 4.3).

Svalbard has three horsetails: field horsetail (*Equisetum arvense*), dwarf horsetail (*E. scirpoides*), and variegated horsetail (*E. variegatum*). The stem of horsetail is very thin, but rough because it is coated with silicates. The silicate-coated horsetail used to be used as an abradant to clean metal cooking pots. Horsetails are also called as puzzle grass of snake grass. The nickname of horsetails is a "living fossil" because the fossils of their relatives go back 100 million years ago. Diverse horsetails were dominant at that time, and even growing up to 30 m tall (Fig. 4.4).

There are two moonworts in Svalbard, moonwort (*Botrychium lunaria*) and Northern moonwort (*B. boreale*), and both species are critically endangered plants (Table 4.1). When we visit the Ossian Sars-fjellet and Blomstrandøya, just on the overside of Ny-Alesund, we can find Brittle bladder-fern (*Cystopteris fragilis*) and Smooth Woodsia (Woodsia glabella). Smooth Woodsia is an endangered species (Table 4.1).

Fig. 4.4 Field horsetail (*Equisetum arvense*)

When we walk along the seashore, we can find orange, green, and blue green spots on the rocks. They are not traces of paint, but of life: lichens. Lichen is not a single organism but a composite of fungi and algae. Scientists recently began to consider bacteria as one of main members of lichen. The shape of lichen is diverse. Some lichens look like plant leaves, others filaments. Some lichens covering the surface of rocks look like paint peeling off while others look like spots. The common names of some lichens are confused with moss. For instance, lichens with names like "Iceland moss" or "reindeer mosses." But a lichen is neither a moss nor a plant.

A lichen is an important producer in tundra ecosystem. Reindeer lichen, which used to be called as reindeer moss (*Cladonia rangiferina*), is one of the favorite foods for reindeer or caribou in Arctic tundra. Some scientists collected the feces of Svalbard reindeers scattered around Ny-Alesund and extracted the DNA from the feces to identify what kind of food they consumed using DNA sequencing. The feces DNA revealed that the Svalbard reindeer fed on lichens (*Stereocaulon* sp. and *Ochrolechia* sp.) and plant species (*Salix polaris* and *Saxifraga oppositifolia*).

Lichens are also pioneering organisms in glacial retreat areas covered with pebbles and soil that lacks nutrients (Fig. 4.5). They can use atmospheric nitrogen as a nitrogen source, which result in fertilization of the bare soil. Lichen can be a habitat for water bears (Tardigrade) as well as tundra plants that grow after the lichen.

Fig. 4.5 Lichen *Xanthoria*
sp. settled on the rocks near
seashore

Pioneers: Purple Saxifrage

When the glacier covering the land retreats, a new plot of ground surfaces. This
glacial foreland is too barren for any plants to live (Fig. 4.6). The plants that grow on
the barren habitat first are called pioneering plants. Purple saxifrage (*Saxifraga
oppositifolia*) is one of the pioneering plants in Svalbard. Purple saxifrage expands
its habitat into newly exposed areas after deglaciation.

Purple saxifrage is a perennial plant with a very small size (Fig. 4.7). It grows
close to the ground surface in the Arctic tundra. Therefore, it is difficult to measure
its height. Purple saxifrage has tiny pink, purple pink, or reddish violet flowers.
White flowers bloom very rarely. Purple saxifrage has five petals. Alaskan Inuit said
the petals can be eaten. In 2000, the Legislative Assembly of Nunavut, the newest
territory to join Canada the year before, chose the purple saxifrage as the floral
emblem of its territory. They make tea using the stems and leaves of purple
saxifrage. I would love to enjoy a cup of purple saxifrage tea.

Even though the name of the plant is "purple" saxifrage, the color of the flowers
are not exactly purple. It is pink, purple pink, or reddish violet. Sometimes I was
lucky to find white flowers. Purple saxifrage has two pistils fusing at base, which
looks like a jester's hat.

Purple saxifrage leaves are thick, orbicular, or triangle shaped. Because the leaves
have no stalk, they are attached directly to the stem. They look more like scale of a
fish than leaves. Purple saxifrage leaves arrange themselves opposite from each

Fig. 4.6 Barren glacial foreland

Fig. 4.7 Purple saxifrage

other. The scientific name of purple saxifrage originated from the array of leaves; opposit (opposite) + folia (leaf).

Purple saxifrage can live in the Arctic polar desert, where it has little water due to low precipitation and low soil moisture content. In the dry environment, they endure stress due to water shortage. The leaves of purple saxifrage can conduct photosynthesis, grow, and even reproduction in the adverse environmental conditions. The time of growth and flowering is very short in the Arctic. When will spring begin? Purple saxifrage is a diligent plant. It blooms early in spring. Purple saxifrage has female and male reproductive organs in its flower. A flower of purple saxifrage has both pistils and stamens. The female and male reproductive cells of purple saxifrage can fertilize and make seeds. When scientists did self-pollination manually in a flower, purple saxifrage had some seeds. Purple saxifrage is self-compatible. Self-fertilization, however, is known to suppress genetic diversity in the next generation. Purple saxifrage wisely avoids self-fertilization by using time difference of maturing the reproductive organs. The anthers of purple saxifrage usually are not ready to spread pollens when the stigmas are waiting to meet pollens. At the end of flowering, the anthers touch the stigmas, but it is too late. The stigmas would have already finished their fertilization with other faster pollens.

Purple saxifrage grows throughout the Arctic. Two subspecies have been proposed for Arctic purple saxifrage on the basis of their distribution area: *S. oppositifolia* ssp. *oppositifolia* and ssp. *glandulisepala.* Their distribution do not overlap. Subspecies *oppositifolia* is distributed on central Siberia, northern Europe, and Greenland. On the contrary, subspecies *glandulisepala* cover northeast Siberia, North America, and northern Greenland. This grouping of purple saxifrage was supported by a genetic marker, *psb*A–*trn*H intergenic spacer of chloroplast DNA (cpDNA). cpDNA possesses more than 100 genes, and most of them participate in photosynthesis. One of the genes is *psb*A, the function of which is involved in electron transportation during photosynthesis. The *psb*A gene is followed by *trn*H gene, and non-transcribed sequences are located between the two genes. The intergenic non-transcribed region is called as *trn*H–*psb*A spacer. Although the *trn*H–*psb*A spacer is short (approximately 450 base pairs), it is the most variable region in cpDNA and is easily amplified in almost land plants. Thus, the *trn*H–*psb*A spacer is used as a DNA barcode to identify plant species.

Dr. Rolf Holderegger, a postdoc at the laboratory of Richard Abbott, a professor at the University of St Andrews, the oldest university in Scotland, analyzed the *psb*A–*trn*H intergenic spacer of purple saxifrage. They found two main haplotypes with similar geographical distributions with the two subspecies: East Asian-North American haplotype and Eurasian haplotype of the *psb*A–*trn*H spacer. Eurasian haplotype has 17 additive bases inside the psbA–trnH spacer, while the East Asian-North American haplotypes does not have them. There is, however, little morphological difference between those two subspecies. The only difference is glandular hairs on sepals that subspecies *glandulisepala* has. The proposal of the two subspecies has not been officially accepted. Therefore, I would like to give an account of purple saxifrage including the two subspecies as a whole. We need to check whether the two subspecies is able to re-produce or not.

Purple saxifrage has two different growth forms: prostrate and cushion types (Fig. 4.8). At first, it was confusing whether the two types were of the same species or different. There was a proposal that the prostrate type should be distinguished from the cushion type by their morphological difference. That is, the prostrate type has leaves with cilia margin, and open flowers with narrow and nonoverlapping petals, whereas the cushion type has leaves without cilia, and urn-shaped flowers with broad and overlapping petals. It was reported that the cushion type is common on snow-free ridges and in early flowering, and the prostrate

Fig. 4.8 Purple saxifrage with prostrate type (left) and cushion types (right)

type is common on areas with prolonged snow cover and late flowering. The length of internode between leaves in a shoot is different. Cushion type has a short internode, and prostrate type has a long internode. The prostrate shoot with long internode shoots can root more easily than the cushion type. Thus, prostrate type seemed to absorb water and nutrient effectively and be robust enough to withstand being runoff along melting ice water on the glacial retreat area. With the same dry weight, the prostrate type can cover a larger area, but the flower number is smaller than the cushion type. It seemed that the prostrate type expands on the newly open glacial retreat area but allocates less resource for sexual reproduction. Therefore, it can be said that the prostrate-type purple saxifrage "wins," and is the pioneer plant on the glacial retreat plain. Individual plants of the cushion type stand on their shoulders. Because they are densely concentrated, they are well tolerated to wind and drying. They produce more seeds, thus they seem to contribute more to the survival and reproduction of the purple saxifrage. The morphological variation, however, was continuous, and they are not separated into distinct characters. Both types, however, can be found at the same area. Still two growth forms are considered as a species.

Some scientists have suggested that the prostrate type and cushion type are determined by the difference of diploid and tetraploid. The inconsistency of the ploidy level would prevent reproduction between the two types, therefore, they proposed subspecies for the two growth types. A suggestion was proposed that the

Fig. 4.9 Hypothesis for
plant form determination of
purple saxifrage

2N	4N
Aa x Aa	Aaaa x Aaaa
(Cushion x Cushion)	(Prostrate x Prostrate)
AA 2Aa aa	9AAaa 6Aaaa aaaa
Die Cushion Prostrate	Die Prostrate Prostrate

cushion type is diploid while the prostrate type is tetraploid. There had been no direct evidence to show the relationship of ploidy level and growth type. The chromosome numbers of purple saxifrage were rarely identified. Several references have told that diploid ($2n = 26$) and tetraploid ($2n = 52$) plants were living in Svalbard. Karl Flovik had counted as many as 52 chromosomes of the purple saxifrage collected from Isfjorden near Longyearbyen. The Japanese expert for another chromosome study, Tsuneo Funamoto, however, identified 28 chromosomes from the purple saxifrage collected from Altai, Russia. They used a classic method: they had stained plant tissues, carried out slide preparation, and observed chromosomes using the microscope. Pernille Eidesen at University Centre in Svalbard and her colleagues applied a more simple technique to identify the ploidy level. They used flow cytometry to measure the fluorescence intensities of the plant cells. Even though they could not count the exact chromosome numbers, they could distinguish diploid from tetraploid plants. They found that tetraploid purple saxifrage is prostrate type, while the diploid plants showed both prostrate and cushion types. They also observed that the tetraploid plants are more frequent in slope with a narrower niche, but the diploid plants are more frequent in ridge with less snow cover, which usually melts out early in spring.

I would like to suggest that the growth type of this plant would be determined by several genes (Fig. 4.9). Let us name the gene as growth form determining (*gfd*) gene of purple saxifrage. Let us presume homozygous recessive *gfd* gene results in the prostrate type and heterozygous the cushion type. Then what about homozygous dominant *gfd* gene? If the plants with homozygous dominant *gfd* gene die, and the dominancy of the prostrate and cushion is almost the same, in which case we then get almost the same number of plants of prostrate and cushion type. The 100 years of investigation is not yet over.

Taking Care of Babies: Alpine Bistort

There has been some confusion around the scientific name of alpine bistort. In 1753, Carl von Linné had categorized alpine bistort into genus *Polygonum* in his wonderful book, "Species Plantarum," and the first scientific name of this plant was *Polygonum viviparum*. Seven years later, the French botanist Antoine Delarbre moved *Polygonum viviparum* to genus *Bistorta*, and it took on a new name, *Bistorta vivipara*. Over 200 years later in 1988, Louis Ronse De Craene and John Akeroyd insisted that this plant belonged to genus *Persicaria* on the basis of the morphology

Fig. 4.10 The young
flowering stalks of Arctic
bistort with bulbils

of its flowers, and *Persicaria vivipara* was accepted as its name in the Plant List
(http://www.theplantlist.org/tpl1.1/record/kew-2572250). The flora of Svalbard,
however, chose *Bistorta vivipara* as the scientific name of the plant.

The phylogenetic analysis on the basis of DNA sequences such as internal
transcribed spacer (ITS), and chloroplast gene for ribulose-1,5-bisphosphate carbox-
ylase/oxygenase large subunit (rbcL) showed that this plant is closely related to
genus *Bistorta*. The flora of Svalbard seemed to have made the right choice in
naming alpine bistort. *Bistorta vivipara* living in Arctic and Alpine, however, has
at least 15 ecotypes. The ecotypes have been identified by microsatellite loci or
chloroplast DNA spacer sequences. We should check the ITS sequences of the
ecotypes. More studies were needed to elucidate the phylogenetic relationship of
Bistorta vivipara and other plants in genera *Polygonum* or *Persicaria*. On the field,
we can identify Arctic bistort by its flowers (Fig. 4.10). At first sight, the white or
pink flowers bloom on upper part of one stem, and it looks like two or more leaves
come out from the stem. In fact, two or three leaves emerge directly from a rhizome,
which is a thick root. The inflorescence stalk also arise from the same rhizome. The
leaves emerge above ground after enduring the "preformation" under the soil. For
3 years, Arctic bistort produces leaf primordium, and it extends leaves above the
ground in the fourth year. From June, right after the melting of snow, two or three
leaves emerge out of the ground mature rapidly. They conduct photosynthesis and
senesce by Arctic autumn around mid-August. Five to six leaf primordia are still
waiting their turns under the ground.

The flowers of Arctic bistort are too tiny to observe their structure by our naked
eye. Each flower is smaller than 5 mm in size. Sometimes two or three white stigmas

Fig. 4.11 Flowers of Arctic bistort. Two or three white stigmas spring out of the flower

resting on the styles spring out of the flower (Fig. 4.11). These tiny flowers cover the upper part of the inflorescence stalk. Just under the white or pink flowers, the red or brown bulbils embrace the lower part of the inflorescence stalk. On the one inflorescence, Arctic bistort produces asexual bulbils as well as sexual flowers (Fig. 4.10).

The ratio of flowers and bulbils is different from plant to plant. Even some inflorescence produces bulbils only. In fact, the flowers and bulbils of Arctic bistort play seesaw game. Flowers cannot be increased without reducing bulbils and vice versa. It seemed that Arctic bistort flowers prefer adequate growing period. If the growing season is short, the inflorescence has fewer flowers and more bulbils. And, the short growing season resulted in the decrease of the bulbils fresh weight. The increasing temperatures led to increase in the number and weight of bulbils. The starch of the rhizome seemed to be consumed for the production of inflorescence. Therefore, the numbers of flowers and bulbils depend on the biomass of rhizome as well as genetic variations and environmental conditions.

Arctic bistorts take care of their babies under the ground. They overcome the short growing season in the Arctic region by preparing baby plants under the frozen ground. The inflorescence stalk is also produced under the ground. The developmental process from the inflorescence primordium to mature flowers and bulbils takes 4 years. As soon as they emerge from the ground, the vegetative and reproductive structures grow vigorously showing their dynamic life power.

The bulbils that fall to the ground can grow new plants. The new plants grown from the bulbils are "clones" while the bulbil-producing plant is their "parent." The clones are genetically identical to the parent.

Arctic bistort bulbils are one of the favorite foods of rock ptarmigans (*Lagopus mutus*). The bulbils are rich in starch, protein, and phosphorus. The main food for chicks younger than 25 days is bulbils of Arctic bistort. The bulbils seem to be the most important food for rock ptarmigan's chicks during their first few weeks after their birth. Arctic bistort nourishes the babies of rock ptarmigans by supplying the bulbils as food.

Gender Issue: Willows

A complete flower is bisexual, containing both female and male reproductive structure. Some plants, however, have only one sex in an individual organism. Therefore, there are male or female plants, having male or female flowers, respectively. We call this dioecious plant, which has distinct male and female organisms.

Willows are such dioecious plants. Willows in temperate region are tall trees, water-rich areas such as stream banks, beside a brook or creek. Willows living in the Arctic are not trees but creeping shrubs, still they love moist soil. In Svalbard, five willows have been reported: polar willow (*Salix polaris*), dwarf willow (*Salix herbacea*), net-leaved willow (*Salix reticulata*), woolly willow (*Salix lanata*), and Arctic willow (*Salix arctica*).

The Arctic willow appeared in a well-known book that surveyed Svalbard plants, written by Olaf Rønning and published in 1996 with the title, "The Flora of Svalbard." Olaf Rønning described that Arctic willow (tundra willow) was "only found in inner Isfjorden and in Kongsfjorden. The species has died at the only known locality in Kongsfjorden and is severely threatened at the other known locality in Adventdalen." The rare willow may have already been extinct in Svalbard. It does not appear on the website "The Flora of Svalbard." When I have observed Arctic willow in Zackenberg on Eastern Greenland, it looks larger than polar willow in plant size. The young leaves of Arctic willow covered with white hairs, but those of polar willow did not.

The woolly willow seems to be close to extinction in Svalbard. Male plants only have been found in Svalbard. The website "The Flora of Svalbard" revealed that "one shrub close to the air strip in Ny-Ålesund ... was grazed to death by reindeer." Four male plants of woolly willow, probably clonal, were discovered in 1963 in Adventdalen. According to "The Flora of Svalbard," they were "living within an area of 10×30 m^2, in a seepage on a terrace brink faced by erosion. The plant is acutely endangered and will disappear within short time."

The net-leaved willow is distinguished from other willows in that they have net-shaped and depressed leaf veins. The species name "*reticulata*" of this plant came from the shape of reticulate leaf surface. The underside of the leaf is covered with fine silky hairs. Thus, net-leaved willow has two colors of leaves: the dark green on the upper side and nearly white on the underside.

The dwarf willow forms mat, where tiny leaves of dwarf willow appear to rise from the ground. In reality, leaf-bearing shoots ascend from the branches stranding out from subterranean stems. I have never seen this plant. It has only been reported from Sørkapp Land and Wedel Jarlsberg Land on Southwest Spitsbergen and on

Fig. 4.12 Male plants of polar willow. The red stamens are carrying yellow pollen

Bjørnøya. The hybrids between dwarf willow and polar willow were reported, leaves of which showed intermediate shape.

The polar willow is also a mat-forming prostrate shrub. Fortunately, it is common throughout Svalbard. Polar willows, like other willows, have separate female and male individuals (Fig. 4.12). Willow flowers have no petals. So at first glance, it does not look like a flower but like a tiny ball of soft fur. The flower of willows are called "catkin," which is a slim oval or cylinder shape. The female catkin of a willow is covered with numerous ovaries, which have reddish styles and stigmas. The male catkin of willow produce red stamens carrying yellow pollen. After fertilization, female willows produce dark reddish capsules, and each capsule contains numerous seeds with white long hairs (Fig. 4.13). When the capsule opens, white fluffs are blown in the wind and the willow seeds are dispersed (Fig. 4.14).

How is the sex of a willow determined? Willows have sex chromosomes. Linkage mapping of the markers on the willow genome gives clues for the sex determination of willow. The basic chromosome number of willows is 19, that is, a haploid gamete has 19 chromosomes ($x = 19$). The net-leaved willow is diploid ($2n = 38$), and polar willow is tetraploid ($2n = 76$). *Salix suchowensis* has one sex chromosome, and the sex is determined by ZW system. In the case of *Homo sapiens*, sex is determined by XY system. Women have XX chromosomes and men have XY chromosomes. Women have the same type of chromosomes, so they are called to be of homogametic sex. On the other hand, men have different type of sex chromosomes, hence they are called be of heterogametic sex. In the case of *Salix suchowensis*, chromosome XV (15th) is different between female and male. The female plants have different types of chromosomes, namely ZW while male plants have the same type, ZZ. Therefore, the female of *Salix suchowensis* is of heterogametic sex.

Fig. 4.13 Female polar willow after fertilization

Fig. 4.14 Red capsules of female catkins carrying many seeds (left), a seed of the polar willow with hairy appendages (right)

Robert Crawford and JEAN Balfour have counted the numbers of female and male willow plants. They found that the polar willow in Spitsbergen, dwarf willow in Iceland, Arctic willow in Canadian High Arctic and northern Greenland, and glaucous willow (*S. glauca*) in Western Greenland showed similar female-biased sex ratios. There were more female plants than male plants (ca. 60:40). They could not find any morphological differences in the leaves of male and female willows. The female dominant sex ratio has been reported in basket willow (*Salix viminalis*) living in Sweden, too. Swedish scientists found a locus on sex chromosome XV related sex ratio distorter of the basket willow and made a model of allelic incompatibility of the sex ratio distorter locus. When the specific DNA sequences (allele) of the sex ratio distorter locus, the allele of which inherited from parents meet together, the male zygote with both ZZ chromosomes cannot grow into a mature seed. Therefore, the number of male plants is less than that of female plants. What is better about more females? It is possible that the excess of female plants results in more seed production.

By the way, the chromosome size and genome structure of the sex chromosomes Z and W of basket willow are almost the same. Even though basket willow has homomorphic sex chromosomes, the expression of genes of the sex chromosomes is slightly different between female and male plants. The genes of Z chromosome are slightly more expressed than W chromosome. The male-biased expression occurs in reproductive tissue, and this is called "the masculinization of gene expression." In conclusion, the differences in gene expression of the sex chromosomes, lead to sexual differentiation of basket willow.

The Glory of the Past: Mountain Avens

A big project to secure the longest ice cores started at the farthest place in the world in 1996. The European Project for Ice Coring in Antarctica (EPICA) set out to drill ice cores at Dome Concordia, which is located in the high ice pack on 3233 m above sea level. The European Science Foundation had boldly supported this 10-year project. More than 50 scientists joined this project to secure the ice cores and to analyze accurately very small amounts of substances. They could analyze a complete Antarctic climate record over the past 800,000 years. European scientist had already had data on the past climate of the Northern Hemisphere. These data came from Greenland; GRIP (North Greenland Ice Core Project), and NGRIP (North Greenland Ice Core Project). The Arctic and Antarctic data matched each other. They have confirmed from the data that there have been several ice ages on Earth. The recent ice age is called the Last Glacial Period (LGP), which had started about 115,000 years ago and ended just 11,700 years ago. Among the LGP the latest frozen season is the Younger Dryas (ca. 12,900 to ca. 11,700 years BP) which was named after a cold-tolerant plant with beautiful flowers, mountain avens (*Dryas octopetala*).

The ice coring data from both the Arctic (GRIP and NGRIP) and Antarctic data (EPICA) showed a sharp falling in temperature in the Younger Dryas. The ice cores taken from Greenland provide the past records of the cold event during the Younger Dryas. The isotopic ratio of oxygen decreased by 3‰, which means 9 °C of cooling during the Younger Dryas. The nitrogen isotope in the fine air bubbles trapped in the ice indicated that the temperature of northern Greenland was 15 °C lower than that of the present. During the Younger Dryas, Europe had also experienced a colder and/or drier climate. The ice sheet covered the Scandinavian peninsular, and mountain glaciers advanced on the Alps (Figs. 4.15 and 4.16).

The name of Younger Dryas was not given by the glaciologists. The name was given by ecologist some time ago. Johannes Iversen, a Danish plant ecologist demonstrated species composition of the late glacial flora using pollen study. Denmark has a strong history of paleoecology using plant pollen. Nikolaj Hartz and his colleague Vilhelm Milther analyzed flora and fauna in deposits at Allerød

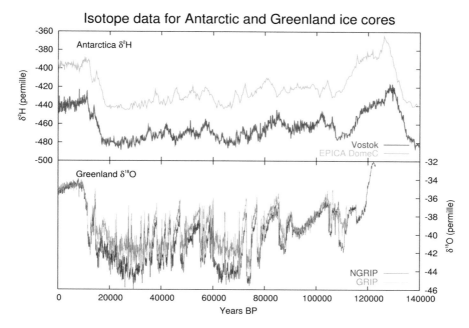

Fig. 4.15 Vostok and EPICA Dome C data from Antarctica, and NGRIP and GRIP data from Greenland showing the past climatic records. (© Leland McInnes, available via license: CC BY 4.0)

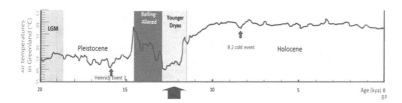

Fig. 4.16 The air temperatures in Greenland show the Younger Dryas about 12,000 years ago. (© Paataliputr; original data from Scientific Reports 7: 40338 (2017), available via license: CC BY-SA 4.0)

through the last ice age. They found leaves of mountain avens in the sand layers more than 100 years ago (Fig. 4.17). Allerød is an archaeologically important small town near Copenhagen. Iversen had used the term "Younger Dryas Time" in his paper published in 1942.

Mountain avens has unique leaves which are readily identified (Fig. 4.18). The leaves are rugged with uneven margins. They resemble an oak leaf, and the meaning of the genus name *Dryas* is 'oak nymph'. Thanks to the identifiable leaves, scientists easily proved its presence in the sediment, and found that mountain avens was living in northern Europe during the ice age. Even though this plant is not the number one

Fig. 4.17 Mature flowers and seeds of mountain avens. (Courtesy of Youngsim Hwang)

Fig. 4.18 A flower and leaves of mountain avens

plant in Northern hemisphere in the Ice age, the Younger Dryas has been given its name after the mountain avens (*Dryas octopetala*). In fact, during the Younger Dryas, mountain avens was not the only plant that endured the harsh ice age. It had friends plants such as polar willow (*Salix polaris*), dwarf birch (*Betula nana*), lesser clubmoss (*Selaginella selaginoides*), and club moss (*Lycopodium selago*).

Still, mountain avens is representative species in frozen world. It is the most important plant in terms of biomass, together with willows and sedges in Eurasian tundras. Mountain avens plays a role in the tundra ecosystem. When a land ice disappears, new barren area is revealed following glacier retreat. The barren area is occupied by microbes and plants through succession. One of the plants is mountain avens. The previous margin of glacier is usually marked with a pile of debris, which is called as moraine. Some mountain avens settle down on the moraine slopes, and the plant slows the flow of water and particles, and stabilizes the debris of glacier. This change in geomorphic process makes possible the establishment of other plants on the glacier retreat area. Therefore, mountain avens is sometimes called as an ecosystem engineer species.

Mountain avens has an elegant flower that does not fit the image of engineer. The flower has eight white petals which are relatively large, compared to other arctic plants. It is a heliotropic flower, which follows the movement of the sun like a sunflower. When the flower is looking at the sun, the temperature inside the flower is more than 3 °C higher than when it is not. The seed weight of plants with petal is 45% heavier than that of plants without the petal. The heliotropic movement of this flower seems to be beneficial for this plant.

The Younger Dryas coincides with the beginning of the Neolithic. Archeologist think that the climatic changes during the Younger Dryas have forced human beings to develop farming. Long ago, our ancestors who lived in Northern Europe might have been comforted by the beautiful white flowers while working hard on the field.

The Facilitator or Nursing Plant: Moss Campion

Several decades before mountain avens' growth in the glacial retreat, some pioneer plants settled down in the uncharted land without vegetation. Moss campion (*Silene acaulis*) is one of the pioneer plants. How can moss campion survive in the rarely vegetated habitat? It is probably due to the distinctive form of the plant.

Moss campion is a cushion plant, which forms a low-growing round mat (Fig. 4.19). Moss campion cushion looks like a dome. Numerous stems and leaves densely covering the stems form the dome-shaped cushion. Numerous compact stems developed from underground branching stems (caudex), which are connected to a deep taproot. Because tiny flat and linear leaves with acute apex overlap each other, there is no gap for even one needle on the surface of the moss campion cushion. Each cushion consists of an individual moss campion.

These characteristics of moss campion are good at surviving in harsh environments such as high mountains and arctic tundra. Glacier retreat is a representative environment where this cushion plant enters as a pioneer. When the glacial ice melting is more than snowfall accumulation, the glacier end-line retreats and land beneath the glacier is revealed. Temperature increasing and snowfall decreasing have caused retreat in almost every glacier in Svalbard. Because soil formed yet in the glacier retreat area, it is a nutrient poor and water-limited place in low temperatures. With tightly packed stems, moss campion is able to trap heat. The temperature inside the moss campion cushion is considerably higher than that of outside air. Thanks to the small and fleshy leaves, which reduce transpiration of the plant, moss campion can retain moisture and water in dry environment. The dome-shaped cushion can also reduce evaporation by slowing down airflow over the plant. The

Fig. 4.19 Moss campion

long taproot can absorb and reserve water effectively in the well-drained environment.

Moss campion can help the growth of other plants coming after the primary succession. So it is called as facilitator or nurse plant. Alpine bistort (*Bistorta vivipara*), for example, has significantly bigger in moss campion cushions compared to outside (Fig. 4.20). Because moss campion makes better microclimatic conditions by increasing temperature, retaining moisture, and trapping litter, plants living inside the cushion plant can grow in a slightly favorable microenvironment. Moss campion possesses abundant plant species compared to outside. Not only plants, but also arthropods were more diverse and abundant in moss campion than surrounding habitats.

Moss campion has two different types of flowers: it has female and bisexual flowers (Fig. 4.21). Some researchers include a male plant, but it seems to be bisexual plants with short styles that functioned as male plants. In Svalbard, the frequency of female moss campion is positively correlated to latitude; the female frequency increases with latitude. Female moss campion seems to be more viable than the bisexual one. The diameter of moss campion plants is proportional to its lifetime. Generally, the diameter of a cushion with female flowers is larger than that of bisexual flowers. Female flowers of moss campion also have high productivity than the bisexual ones. Female plants produce more than four times as many seeds over their lifetime. Why is the female moss campion strong?

To answer this question, we need to investigate multiple causes, such as environmental factors, sex determination mechanism, maternal relationship, fungal infection, and so on. One of the possible answers is the morphological difference

Fig. 4.20 Alpine bistort and Arctic willow are growing in the moss campion cushion

Fig. 4.21 Female flower (left) and bisexual flower (right) of moss campion

in the length of style. Flowers having short styles cannot produce degenerated ovules. The higher percentage in bisexual plants had short styles than that in female plants. Even though bisexual plants with long style produce the same number of normal ovules as female plants, those plants are uncommon. Some moss campion plants are infected by the anther-smut fungus *Microbotryum violaceum*. Smut infection causes sterility in both female and bisexual plants. This infection might result in the gender difference in age and productivity (Fig. 4.21).

Arctic plants must grow and reproduce with limited time and energy under a short growing season and low temperature. Cold and strong wind can cause infertility though pollen limitation. Bisexual moss campion seems to be more favorable than female one, because it can overcome this limited environment through self-pollination. To make lower risk of inbreeding, it might be advantageous to have an alternative reproduction system. The outcrossing through female plants can be an alternative way for moss campion.

Moss campion is a long lived but slow growing plant. The plants with a diameter of 10 cm might be over 50 years old. Infant mortality rate of moss campion is very high. It is not uncommon for these plants to die within the first year of germination. That is not the end. Many seedlings die before becoming adult plants. If a young moss campion exceeds a critical size, the plant will live long enough. "Long life and plenty of offspring to moss campion!"

Live Together with Tiny Guests: Mountain Sorrel

The scientific name of mountain sorrel is *Oxyria digyna*. *Oxyria* means "acidic" and the leaves of the plant have a sour taste due to oxalic acid. The kidney-shaped leaves of mountain sorrel are thick and slightly fleshy (Fig. 4.22). The fact that the taste is known means that someone has eaten the plant: Indigenous people. Cup'ig Eskimo of Nunivak Island begin to harvest the leaves of mountain sorrel (quulistar) as food in early spring. Mountain sorrel leaves have been used as indigenous plants in Yupik (quunartiarraat), Chukchi and Siberian Yupik (k'ugyln'ik), and Inupiaq (qufuliq, qunulliq). Sami has also used mountain sorrel as a dessert Jåbmå. They cook the leaves to a stew and serve with milk and sugar. Mountain sorrel leaves were also boiled to make a pulp, which heated with reindeer milk to make a porridge in traditional Sami food. The vitamin C content is 36 mg per 100 g raw mountain sorrel. If it is boiled for 10 min, vitamin C is reduced to one-seventh. Except for whale epidermis (Muktuk) and raw char, mountain sorrel would have been a major source of vitamin C for the indigenous people.

Like its common name, mountain sorrel grows not only in Arctic such as Greenland, Alaska, and Svalbard, but also on mountains of Northern Hemisphere: Alps in Europe, Rocky mountains in North America, Himalayas in Asia, Western and North eastern areas of China, and the northern mountains of the Korean Peninsula. Because it is distributed widely, the research comparing Arctic and alpine mountain sorrel has been performed since 1940s. Mountain sorrel has two ecotypes, the northern and southern types on the basis of morphological characteristics. The dividing line coincides with the maximum extent of Pleistocene glaciation. The northern type has less flowering branches but more rhizome than the southern type. Both types do not bloom under continuous short days. They produce flowers when they get long daylight. The normal mountain sorrel flowers have six stamens. In North America, unusual flowers bearing two stamens or none are found in the southern type. But the stamen number Svalbard mountain sorrel is variable, some has aborted stamen (Fig. 4.23). The secret of the stamen development is still unknown.

Fig. 4.22 Mountain sorrel

Fig. 4.23 Flowers of
mountain sorrel

    Microbiome is a rising star in modern biology. It is the collective microbes living
in an environmental niche including individual organisms. Microbiomes of humans
are known to even affect the health and personality of the hosts. Plants also have
microbiomes such as endosphere microbiome living inside the plant tissues and
rhizosphere microbiome living around the roots. Microbiome of mountain sorrel has

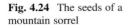

Fig. 4.24 The seeds of a mountain sorrel

been studied by Finnish and Dutch researchers. Riitta Nissinen and Jan Dirk van Elsas at University of Groningen had sowed the seeds of the study, and Manoj Kumar has blossomed the study on the mountain sorrel's microbiome.

They have analyzed the community of potential nitrogen fixing bacteria in the soil, rhizosphere, and endosphere mountain sorrel by using *nifH* gene. The soil and the rhizosphere microbiomes were similar to each other, but the microbiome inside the plant was very different. The major endophytic bacteria living in mountain sorrel belong to genus *Bradyrhizobium*, *Burkholderia*, *Clostridium*, *Caldicellulosiruptor* (with a very strange name), and *Leptothrix*. They also include photosynthetic cyanobacteria, which are able to fix nitrogen: *Nostoc* and *Synechococcus* (another strange name). When they analyzed the microbiome using 16S rDNA gene, several bacterial genus were added to the endophytic bacterial list: *Janthinobacterium*, *Methylibium*, and *Rhodoplanes*. Among the bacteria, *Clostridium* spp. have been observed in the seeds of mountain sorrel, too. Is there any chance that the bacterium is transmitted through the seed to the next generation? (Fig. 4.24)

Overcome Stress: Whitlow-Grasses

The Arctic tundra or Central London, which is better to live? Wherever we live, we face stress every moment to cope with. Sometimes we can go on vacation to avoid stress, but plants cannot. Plants just have to get over the stress. Plants have their own secret to overcome cold stress, and various cold stress-induced genes have been reported. One of them is the *cor15* gene, which means cold-regulated (*cor*) gene.

Whitlow-grass (*Draba*) also has *cor15*, even more than one. Two different sequences of *cor15* genes have been found in alpine whitlow-grass (*D. alpina*). Thale cress (*Arabidopsis thaliana*) also has two homologs of *cor15*, which are aligned in the same chromosome. The two *cor15* genes have different expression patterns in where and when in response to cold stress. This gene encodes a 15-kDa polypeptide COR15. The mature COR15 polypeptide is hydrophilic and boiling stable. COR15 has chloroplast transit peptide to make the protein move to the stroma of chloroplast. In fact, the mature protein of COR15 is present in the soluble fraction of chloroplasts.

The function of *Arabidopsis* COR15 is closely related to the cold acclimation (Fig. 4.25). Cold acclimation is a phenomenon that plants acquire freezing tolerance by being exposed at low, nonfreezing temperature. When we put plants growing in room temperature to freezer, the plants will be killed. But if we lower the temperature and let the plants in 10 °C for several days, the plants can live in a freezer of -20 °C. The mature COR15 is detected only in the cold-acclimated leaves; it is neither detected in leaves without cold-treatment nor cold-acclimated roots. We can find COR15 from the chloroplast solution getting cold stress. This protein seems involved in enduring the dehydration that occurs in cold-stressed plants. In non-acclimated *Arabidopsis*, the function of COR15 is not so impressive. The freezing tolerance of chloroplasts increases by 1~2 °C through overproduction of COR15a. The expression of *cor15* is less affected than the cold acclimation that increases the freezing tolerance of chloroplast by 6 °C. Plants experiencing cold acclimation obtain freezing tolerance by multigene expression. The copy numbers of cold-related genes depends on ploidy levels of the plants.

Whitlow-grass plants living in Svalbard each have different ploidy levels. Austrian draba (*D. fladnizensis*), snow whitlow-grass (*D. nivalis*), and Ellesmere Island whitlow-grass (*D. subcapitata*) are diploid ($2x = 16$). Lapland whitlow-grass (*D. lactea*), Small-flowered draba (*D. micropetala*), rock whitlow-grass (*D. norvegica*), and few-flowered whitlow-grass (*D. pauciflora*) are hexaploid ($6x = 48$). Gray-leaf draba (*D. cinerea*), smooth draba (*D. daurica*), and Gredin's whitlow-grass (*D. oxycarpa*) are octaploid ($8x = 64$). Alpine whitlow-grass (*D. alpina*) and Arctic draba (*D. arctica*) possess five times of chromosomes than diploid plants; they are decaploid ($10x = 80$). Svalbard flattop draba (*D. corymbosa*) even has 128 chromosomes ($16x$). Then how do they have different ploidy levels?

The plant biologists estimate Cretaceous–Paleogene (K–Pg) extinction is closely related to the spreading of polyploid. The unreduced gametes could be produced in response to environmental stress. These diploid gametes are not easy to keep the next generation after fertilizing with normal gametes with haploid genome, because triploid plants are difficult to produce reproductive cell through meiosis. On the other hand, when the diploid gametes fertilized with diploid gametes, it would be easy to keep the next generation. The best way to meet diploid gametes is self-pollination. Because some pollinators disappeared during the extinction event, plants relying on pollinators might be at risk. Thus, self-pollinating plants might survive better than plants relying on pollinators. The frequent self-pollination was more advantageous to diploid gametes to produce tetraploid offspring. Even though it is hard to say what happened in the past, we can imagine that the past precipitous and/or enormous environmental change might cause the polyploidization in the plant kingdom.

Self-pollination is common in whitlow-grasses. Except Gredin's whitlow-grass (*D. oxycarpa*), self-pollination is the main method of producing seeds in whitlow-grass plants. The percentage of seed production was above 50% in most whitlow-

grasses. On the contrary, it was less than 16% in Gredin's whitlow-grass. The
percentage of self-pollinating seed production of diploid Austrian draba
(*D. fladnizensis*) and Ellesmere Island whitlow-grass (*D. subcapitata*) was 92%
and 100%, respectively. This high level of self-pollination in diploid plants may
make the plants genetically depauperate, because the genetic diversity of the self-
pollinating diploid would be lower than that of the polyploid. For this reason,
although it costs more to maintain more chromosomes, whitlow-grasses with various
ploidy levels have existed so far.

The flower color of Svalbard draba is white or yellow (Fig. 4.26). Gredin's
whitlow-grass and few-flowered whitlow-grass have light yellow petals. The petal
of alpine whitlow-grass and flattop draba is yellow. The other Svalbard draba has
white petals.

Fig. 4.26 Svalbard draba plants with yellow flowers and white flowers

Immigration: Buttercups

The size of most Arctic flowers is very small. So we can easily pass by without even Arctic flowers are generally much smaller than other plants, and so we can easily pass by them without noticing their existence. Buttercups, however, are exceptional and can be easily spotted because they have large flowers (Fig. 4.27). The yellow petals are usually over 1 cm, with clear and defined the male and female gameto- phytes. In the center of the flower, a lot of yellow stamens surround a green ball- shaped structure that is a collection of carpels. A carpel is the female part of the flower, which will eventually develop into fruit and seeds.

Among the 12 buttercups species found in Svalbard, polar kidney buttercup (*Ranunculus wilanderi*) is the rarest buttercup in the world as there are less than 100 plants. Polar kidney buttercup has been reported from only Kapp Thordsen in Dickson Land of Svalbard. This is an endemic plant living in Svalbard only, and the microspecies of the *Ranunculus auricomus* are known to be complex. All members of this complex possess the ability to produce seeds asexually, where the polar kidney buttercup does so by fragmentation of caudex. Caudex of polar kidney buttercup is a kind of branch from the root. This means that the genetic diversity of the polar kidney buttercup population is very low making it vulnerable to environmental changes. In fact, the population consists of only two genotypes.

The small population of polar kidney buttercup has been found in damp moss tundra. If the drainage conditions of the habitat changes and it is turned into a drier environment, polar kidney buttercup could be in danger of extinction. It is exposed to be reindeer and other herbivores as foods. The increase of temperature bring

Fig. 4.27 Buttercup with yellow stamens and ball-shaped carpel

competition to the polar kidney buttercup with other faster growing, fertile species that are migrating from the subarctic region. To avoid the extinction of polar kidney buttercup, they are being kept in botanical gardens such as the Botanical Garden in Natural History Museum, University of Oslo and Tromsø Arctic-Alpine Botanic Garden in The Arctic University Museum of Norway. It is called ex situ conservation. The polar kidney buttercup had been able to settle in Svalbard as there was no competitors, but immigration of other plants presents a risk to the polar kidney buttercup in the future.

Pigmy buttercup (*Ranunculus pygmaeus*) is a small buttercup with tiny flowers (Fig. 4.28). It has low genetic diversity in areas under glaciation during the last ice age. Pigmy buttercups, found in Greenland over to Svalbard and Scandinavia, have the same amplified fragment length polymorphism (AFLP) pattern. After the last ice age, one population of pigmy buttercup from Scandinavia might be initial plants colonizing the deglaciated area of Svalbard. Pigmy buttercup is a strong self-fertilizing plant, therefore even single plant can reproduce and multiply.

The dispersal of buttercups depend on historical events such as past geography and climate change, as well as on dispersal vectors such as migrating birds, wind, ocean currents, and sea ice. Some buttercups could have survived the last ice age in local ice-free refugia in different regions. Then a possible hypothesis is that they were transported following stepping stones across islands such as Iceland, Jan Mayen, Hopen, and Bjørnøya. Some buttercups could have moved from Greenland to Svalbard.

Fig. 4.28 Pigmy buttercup (left) and Snow buttercup (right)

Making Flowers: Polar Campion

There are unique flowers in Svalbard that are unforgettable such as the polar campion (*Silene uralensis* ssp. *arctica*) and the arctic white campion (*Silene involucrata* ssp. *furcata*) blossom unlike other flowers (Figs 4.29 and 4.30). The five sepals are fused to form a balloon-shaped calyx at the bottom of the flower, making the flowers look like balloons. The calyx is a pale violet or light green with stripes of dark violet veins whereas the petals of the flower are grayish violet, lilac, or white color. The flowers tell us whether they succeeded in reproduction or not. Polar campion flowers nod in early flowering stage, but after pollination the flowers are erect.

A 3-h drive north of Warsaw, Poland, leads us to an area called Olsztyn, where the University of Warmia and Mazury is located. Unlike other European universities with hundreds of years of history, this new university was founded in 1999. The Faculty of Biology and Biotechnology has operated a greenhouse, where several Arctic and Antarctic plants are growing. Irena Giełwanowska and some researchers have observed microscopic structure of these polar plants including the arctic white campion. Here, they have put some seeds of Arctic white campion on filter paper saturated with distilled water, which was collected in Svalbard. They cultured them in growth chambers, and counted the number of germinated seeds. The seeds were so

Fig. 4.29 Polar campion

small that the experiment for seed counting required a lot of patience. The average seed length and seed width were just 1.05 mm and 1.18 mm, respectively. The germination rate of Arctic white campion seeds was more than 98% between 12 and 20 °C. In another experiment by Norwegian researchers, the germination rate of the seeds cultured indoor was more than 60%, but it dramatically dropped to less than 4.7 ± 1.2% when the seeds were germinated in the Arctic field.

Flowers have a special meaning to human being. Through a bunch of flowers, we convey thanks, congratulations, and love. When we happen to see beautiful flowers, we smile unconsciously. We feel happy with floral scent, except when we have allergies to pollen. We keep special memories by drying petals. But to plants, flowers have different meanings. When flowers bloom, they bear seeds. It is also the way of a finite organism prolonging life, effectively spreading one's genes by leaving off-spring behind. The fruit and seed, the gift of the flower became the raw material of food and medicine for a long history. How can we possibly dislike such a delicate flower?

Flowers are not mere organs of a plant but a special period in the life cycle of a plant, bearing different generations. The pollen that is produced after meiosis is the male gametophyte. Pollen is only a cell with three nuclei, but it plays a critical role in the reproduction. Pollen move to the female gametophyte, propose, and mix with the genetic material of the female nuclei. Flowers are the very places where a great journey for the next generation begins, fertilization. Therefore, it is quite natural for people to wonder how flowers develop.

Fig. 4.30 Arctic white campion

One of the pioneers of flower development research is Elliot Meyerowitz at the California Institute of Technology. Professor Meyerowitz had obtained several *Arabidopsis* mutant plants from a Dutch professor, Maarten Koornneef (Wageningen Agricultural University). The mutant plants had bloomed into abnormal flowers, and kept homozygous by self-pollination. The mutant plants had names for individual mutations according to their incomplete morphology of the flowers such as *agamous-1, apetala2-1, apetala3-1,* and *pistillata-1.* Students of Meyerowitz observed the developmental process of the mutant flowers using microscopy, but the genes were wrapped in a veil. Based on observing mutant flowers, Meyerowitz proposed a very simple idea to explain the development of flowers, the ABC model (Fig. 4.31).

When we look down the flower from above, we can draw four concentric circles known as whorls. The floral organs are arranged on the whorls. Sepals develop at the outermost, and the place is designated as whorl 1. Petals are formed inside the sepals, and it is designated as whorl 2. Like this, stamens produced at whorl 3, and carpels innermost, whorl 4 in turn. How do plants produce different organs along the four whorls? According to the ABC model, the apical meristem where cells divide and differentiate into floral organs, the genes are differentially expressed depending on the concentric circles.

The genes involved in making flowers can be classified into three groups in the ABC model. The A-class genes are engaged in producing sepals at whorl 1. At whorl 2, A- and B-class genes work together to make petals. B- and C-class genes promote

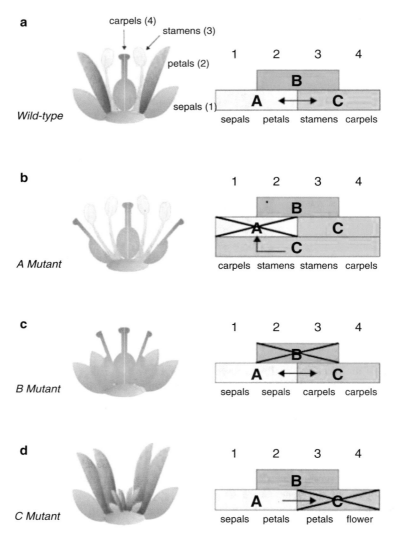

Fig. 4.31 ABC model explaining floral development. (© American Society of Plant Physiologists)

stamen development at whorl 3, and C-class genes carpel development at the innermost whorl 4. But later scientists discovered other genes were involved in flower development. There were some genes acting in ovule development, which were designated as D-class. If this gene breaks down, even though the genes of A-, B-, and C-class are normal, the plant did not produce a normal flower. Since this is called the E group gene, the "ABC model" has been upgraded to "ABCE model." But this is only a model. It was only the result of a genetic study of morphological mutants, but the actual genes were not revealed. They had to find the genes with

DNA sequences belonging to ABC classes. How could scientists find these flower-developing genes at the molecular level?

Hans Sommer and colleagues at Max-Planck-Instituts für Züchtungsforschung (Max Planck Institute for Plant Breeding Research) located in Köln had obtained snapdragon (*Antirrhinum majus*) mutant plants from a British plant geneticist Rosemary Carpenter. These mutant plants were made by transposon mutagenesis, and the breaking genes by transposable elements could be used to identify the DNA sequencing. One of the mutant plants had *deficiens* gene mutation, which caused the changes of petals into sepals and of stamen into carpels. They could isolate the gene using differential screening of the cDNA library, and clone it. The *deficiens* gene belong to homeotic genes regulating the development of specific organs. The *deficiens* gene includes MADS-box, which is a conserved sequence motif translating the DNA-binding domain. The strange name, MADS-box is an acronym referring to previously reported similar homeotic genes originated from the budding yeast (*MCM1*), from the thale cress (*Agamous*), from the snapdragon (*Deficiens*), and from the human (*SRF*). After this finding, scientists had screened other homeotic genes using MADS-box sequences, and linked these genes to the morphological genes of the ABC model, which had been identified by genetic analysis. For example, *apetala-1* (*ap1*) belongs to A-class, *apetala-3* (*ap3*) and *pistillata* (*pi*) to B-class, *agamous* (*ag*) to C-class, and *sepallata-1, -2, -3, -4* (*sep1, 2, 3, 4*) to E-class. Mutant plants of these genes produce unusual flower. For example, petals change to sepals and stamens to carpels in *apefala3-1* mutant plants, in petals to sepals in *pistillata-1* mutant plants, and stamens turn to petals in *agamous-1* mutant plants.

The morphology of these flowers is various, and the ABC(E) model does not perfectly explain the flower development of all plants. Still, many variations of the ABC(E) model is proposed to a specific morphology of flowers, for example, monocots develop floral organs on three whorls instead of four whorls. Over 80 genes have been found to be required in normal floral development and there might be more genes yet to be discovered. Someday, the developmental process of the balloon-like flowers of polar campion and Arctic white campion will be revealed at the molecular level.

Overcoming Drought: Arctic Mouse-Ear

When we walk along the main road in Longyearbyen, we can see the plants with white flowers blooming on the roadside. This plant is so common that it can be easily seen anywhere in Svalbard. The white flower petals are shallowly lobed, and they resemble the shape of a rat's ear, hence the name Arctic mouse-ear (Figs. 4.32 and 4.33).

Arctic mouse-ear (*Cerastium arcticum*) is widely distributed in the Arctic as its name suggests. It is widespread not only in Svalbard but also Franz Joseph Land, Greenland, Northern Iceland, Polar Ural—Novaya Zemlya, and Canadian High Arctic, but it has not been observed in Alaska. The Arctic mouse-ear is found to

Fig. 4.32 Arctic mouse-ear on the Longyearbyen roadside

be well adapted to cold and dry conditions, living in almost all kinds of Arctic environment including polar desert, arctic tundra, and shrub tundra. How could the plant adapt well to the extreme environment of the Arctic? One of its secret turned out to be dehydrin.

Dehydrin is a group of proteins against dehydration. The name of this protein is derived from a combination of **dehydr**ation and prote**in**. Dehydrin is closely related to seed development and protective reactions to cold and drought stress, yet the function of dehydrin has not been fully understood. Dehydration might result in cellular membrane alteration, cytoplasm shrinking, and enzyme activity reduction, etc., which can lead to growth retardation, reproductive reduction, or eventual plant death. Research data shows that dehydrin plays important roles in enzyme cryoprotection and membrane protection in a condition of water shortage or osmotic stress.

Amino acid sequences of dehydrin shows that it is highly hydrophilic. More than 50% of the amino acids of dehydrin have charged or polar residues, over 25% of alanine or glycine, but cysteine and tryptophan are absent. Thus, dehydrin can bind to large quantities of water molecules and ions. Dehydrin has a highly conserved lysine-rich motif, the K-segment, which may be repeated from 1 to 12. Depending on the subgroups of dehydrin, there are two other motifs, the S-segment and the Y-segment. The S-segment seems to promote dehydrin's translocation into the nucleus, and the Y-segment is similar with the nucleotide-binding site of bacterial chaperones. The amino acid sequences between the conserved segments are called as the ϕ-segment, which is not a conserved region but is variable in length. According to the arrangement of the segments, dehydrins are classified into five major groups: K_n, SK_n, K_nS, Y_nK_n, and Y_nSK_n. Most of K_n, SK_n, and K_nS dehydrins are

Fig. 4.33 Arctic mouse-ear forming seeds

synthesized during cold stress, and some of them under desiccation and salt stress. Y_nSK_n dehydrins are produced only by desiccation and salt stress. In the case of Y_nK_n, one subgroup increases under cold stress, and the other subgroup by desiccation and salt stress.

Dehydrins are plant-specific proteins; it seems that all flowering plants possess dehydrins. Among the five groups of dehydrins, K_n and SK_n are also identified in gymnosperm. The Y-segment containing dehydrins are produced in only flowering plants with high levels during seed germination. Dehydrin genes are expressed during seed maturation, but not in well-watered seedlings.

A dehydrin gene has been isolated from Arctic mouse-ear. The gene was named *CarDHN* according to the scientific name *Cerastium arcticum*. The Arctic mouse-ear dehydrin is a small molecule composed of 270 amino acids and belonging to the SK_5, group with one S-segment and five K-segment repeats. The *CarDHN* gene was introduced to a yeast, *Saccharomyces cerevisiae*, and the *CarDHN*-expressing transgenic yeasts acquired stress resistance. The transgenic yeasts could recover more rapidly from stress such as oxidants treatment, high salinity, freezing and thawing, and Zn^{2+}. The transgenic yeasts showed higher fermentation capacity, surviving in glucose-based batch fermentation by producing a higher concentration of alcohol than wild-type yeasts.

The *CarDHN* gene also has been introduced to tobacco. Two dehydrins have already been reported to be found in tobacco, showing different structures compared to the Arctic mouse-ear dehydrin. Although there was no significant difference in morphology between the wild-type and the transgenic plants under drought stress, the transgenic plants showed improved resistance to stress. The *CarDHN*-expressing plants exhibited higher tolerance to salt and osmotic stress during seed germination and seedling growth. The transgenic plants could survive after -10 °C freezing treatment. The *CarDHN* transgenic plants had lower transcription levels of heat shock proteins (HSP70 and HSP26) under drought or salt stress. Heat shock proteins are expressed in various stress conditions as well as heat treatment, therefore the expression of HSP genes can be an indicator of plant response to stresses. The decreased expression of HSP genes in the transgenic tobacco may imply that the *CarDHN* gene enhanced a stress tolerance of the transgenic plants.

What is the role of dehydrins in Arctic plants surviving at the harsh environments? I look forward to uncovering the secret of dehydrins in the Arctic mouse-ear.

Bibliography

Abbott RJ, Comes HP (2004) Evolution in the Arctic: a phylogeographic analysis of the circumarctic plant, *Saxifraga oppositifolia* (Purple saxifrage). New Phytol 161:211–224

Abbott RJ, Smith LC, Milne RI, Crawford RM, Wolff K, Balfour J (2000) Molecular analysis of plant migration and refugia in the Arctic. Science 289(5483):1343–1346

Allagulova CR, Gimalov FR, Shakirova FM, Vakhitov VA (2003) The plant dehydrins: structure and putative functions. Biochemistry (Mosc) 68:945–951

Allen GA, Marr KL, McCormick LJ, Hebda RJ (2012) The impact of Pleistocene climate change on an ancient arctic-alpine plant: multiple lineages of disparate history in *Oxyria digyna*. Ecol Evol 2:649–665

Alsos IG, Westergaard K, Lund L, Sandbakk BE (2004) Floraen i Colesdalen, Svalbard. Blyttia 62:142–150

Alsos IG, Müller E, Eidesen PB (2013) Germinating seeds or bulbils in 87 of 113 tested Arctic species indicate potential for ex situ seed bank storage. Polar Biol 36:819–830

Alström-Rapaport C, Lascoux M, Gullberg U (1997) Sex determination and sex ratio in the dioecious shrub *Salix viminalis* L. Theor Appl Genet 94:493–497

Bauert MR (1996) Genetic diversity and ecotypic differentiation in arctic and alpine populations of *Polygonum viviparum*. Arct Alp Res 28:190–195

Bills JW, Roalson EH, Busch JW, Eidesen PB (2015) Environmental and genetic correlates of allocation to sexual reproduction in the circumpolar plant *Bistorta vivipara*. Am J Bot 102:1174–1186

Birkeland S (2012) Rare to be warm in Svalbard: an ecological and genetic snapshot of four red listed plant species. Master of Science Thesis, UNIS and Universitet i Oslo

Birkeland S, Elisabeth Borgen Skjetne I, Krag Brysting A, Elven R, Greve Alsos I (2017) Living on the edge: conservation genetics of seven thermophilous plant species in a High Arctic archipelago. AoB Plants 9:plx001

Blackburn KB, Harrison JWH (1924) A preliminary account of the chromosomes and chromosome behavior in the Salicaceae. Ann Bot 38:361–378

Blázquez MA, Weigel D (2000) Integration of floral inductive signals in Arabidopsis. Nature 404 (6780):889–892

Bowman JL, Smyth DR, Meyerowitz EM (1989) Genes directing flower development in *Arabidopsis*. Plant Cell 1:37–52

Brochmann C (1993) Reproductive strategies of diploid and polyploid populations of arctic *Draba* (Brassicaceae). Plant Syst Evol 185:55–83

Brysting AK, Gabrielsen TM, Sørlibråten O, Ytrehorn O, Brochmann C (1996) The purple saxifrage, *Saxifraga oppositifolia*, in Svalbard: two taxa or one? Polar Res 15:93–105

Burgess P (ed) (2017) EALLU: indigenous youth, Arctic change & food culture – food, knowledge and how we have thrived on the margins. Arctic Council Sustainable Development Working Group, pp 54

Carlson AE (2013) The younger Dryas climate event. In: Elias SA (ed) The encyclopedia of quaternary science. 3:126–134

Close TJ, Kortt AA, Chandler PM (1989) A cDNA-based comparison of dehydration-induced proteins (dehydrins) in barley and corn. Plant Mol Biol 13:95–108

Coen ES, Romero JM, Doyle S, Elliott R, Murphy G, Carpenter R (1990) Floricaula: a homeotic gene required for flower development in *Antirrhinum majus*. Cell 63:1311–1322

Colombo L, Franken J, Koetje E, Van Went J, Dons HJ, Angenent GC, Van Tunen AJ (1995) The petunia MADS box gene FBP11 determines ovule identity. Plant Cell 7:1859–1868

Crawford RMM, Balfour J (1983) Female predominant sex ratios and physiological differentiation in Arctic willows. J Ecol 71:149–160

Crawford RMM, Balfour J (1990) Female-biased sex ratios and differential growth in Arctic willows. Flora 184:291–302

Dahl E (1937) On the vascular plants of Eastern Svalbard. (Skrifter om Svalbrd og Ishavet Nr. 75). Norway's Svalbard and Arctic Ocean Research Survey, Oslo. 50 p

Darlington CD, Wylie AP (1955) Chromosome atlas of flowering plants. Allen & Unwin, pp 178–179

Diggle P (1997) Extreme preformation in alpine *Polygonum viviparum*: an architectural and developmental analysis. Am J Bot 84:154–169

Eichel J (2019) Vegetation succession and biogeomorphic interactions in Glacier Forelands. In: Heckmann T, Morche D (eds) Geomorphology of Proglacial systems, geography of the physical environment. 19:327–349

Eidesen PB, Müller E, Lettner C, Alsos IG, Bender M, Kristiansen M, Peeters B, Postma F, Verweij KF (2013) Tetraploids do not form cushions: association of ploidy level, growth form and ecology in the High Arctic *Saxifraga oppositifolia* L. s. lat. (Saxifragaceae) in Svalbard. Polar Res 32:1–12

Elven R, Elvebakk A (1996) Part 1. Vascular plants. In Elvebakk A, Prestrud P (eds) A catalogue of Svalbard plants, fungi, algae, and cyanobacteria. Norsk Polarinstitutt Skrifter, pp 9–55, 198

Evju M, Blumentrath S, Hagen D (2010) Nordaust-Svalbard og Søraust-Svalbard naturreservater. Kunnskapsstatus for flora og vegetasjon NINA Rapport 554. (in Norwegian)

Frisvoll AA, Blom HH (1993) Trua moser i Norge med Svalbard; raud liste. - NINA Utredning 42:1–55

Frisvoll AA, Elvebakk A (1996) Bryophytes. In: Elvebakk A, Prestrund P (eds) A catalogue of Svalbard plants, fungi, algae and cyanobacteria. Part 2. Norsk Polarinstitutt, Oslo, pp 57–172 (Norsk Polarinstitut Skrifter; vol 198)

Funamoto T (2010) Somatic chromosomes of ten species of *Saxifraga* L. (Saxifragaceae) in Asian continent, Russia and Mongolia. Chromosome Bot 5:5–13

Galasso G, Banfi E, De Mattia F, Grassi F, Sgorbati S, Labra M (2009) Molecular phylogeny of *Polygonum* L. s.l. (Polygonoideae, Polygona ceae), focusing on European taxa: preliminary results and systematic considerations based on *rbc*L plastidial sequence data. Atti della Società italiana di Scienze naturali e del Museo civico di Storia naturale in Milano 150:113–148

Garbary D, Renzaglia KS, Duckett JG (1993) The phylogeny of land plants: a cladistic analysis based on male gametogenesis. Plant Syst Evol 188:237–269

Geraci JR, Smith TG (1979) Vitamin C in the diet of Inuit hunters from Holman, Northwest Territories. Arctic 32:135–139

Gjoerevoll O, Ronning O (1980) Flowers of Svalbard. Universitetsforlaget, pp 5–7

Grundt HH, Kjølner S, Borgen L, Rieseberg LH, Brochmann C (2006) High biological species diversity in the arctic flora. Proc Natl Acad Sci U S A 103:972–975

Gugerli F, Bauert MR (2001) Growth and reproduction of *Polygonum viviparum* show weak responses to experimentally increased temperature at a Swiss Alpine site. Bot Helv 111 (2):169–180

Hanley S, Barker A, Van Ooijen W, Aldam C, Harris L, Åhman I, Larsson S, Karp A (2002) A genetic linkage map of willow (*Salix viminalis*) based on AFLP and microsatellite markers. Theor Appl Genet 105:1087–1096

Hartz N, Milthers V (1901) Det senglacie ler i Allerød tegelværksgrav. Meddelelser Dansk Geologisk Foreningen 8:31–60

Hedderson TA, Chapman RL, Rootes WL (1996) Phylogenetic relationships of bryophytes inferred from nuclear-encoded rRNA gene sequences. Plant Syst Evol 200:213–224

Heide OM (2005) Ecotypic variation among European Arctic and alpine populations of *Oxyria digyna*. Arct Antarct Alp Res 37:233–238

Hermanutz LA, Innes DJ (1994) Gender variation in *Silene acaulis* (Caryophyllaceae). Plant Syst Evol 191:69–81

Hill W, Jin X-L, Zhang X-H (2016) Expression of an arctic chickweed dehydrin, CarDHN, enhances tolerance to abiotic stress in tobacco plants. Plant Growth Regul 80:323–334

Hoffmann MH, von Hagen KB, Hörandl E, Röser M, Tkach NV (2010) Sources of the arctic flora: origins of arctic species in Ranunculus and related genera. Int J Plant Sci 171:90–106

Holderegger R, Abbott RJ (2003) Phylogeography of the Arctic-alpine *Saxifraga oppositifolia* (Saxifragaceae) and some related taxa based on cpDNA and ITS sequence variation. Am J Bot 90:931–936

Hou J, Ye N, Zhang D, Chen Y, Fang L, Dai X, Yin T (2015) Different autosomes evolved into sex chromosomes in the sister genera of *Salix* and *Populus*. Sci Rep 5:9076

http://panarcticflora.org/results?biogeographic=&bioclimatic=&region=&name=Cerastium +arcticum#paf-420207

http://svalbardflora.no/index.php?id=613

http://www.assembly.nu.ca/about-legislative-assembly/official-flower-nunavut

http://www.luontoportti.com/suomi/en/kukkakasvit/purple-saxifrage

http://www.samer.se/4591

https://en.wikipedia.org/wiki/ABC_model_of_flower_development#References

https://en.wikipedia.org/wiki/List_of_Sami_dishes

https://en.wikipedia.org/wiki/Saxifraga_oppositifolia

https://plants.usda.gov/core/profile?symbol=CEAR16

https://www.canada.ca/en/canadian-heritage/services/provincial-territorial-symbols-canada/nuna vut.html#a3

Hubbard A, Andreassen K, Auriac A, Whitehouse PL, Stroeven A, Shackleton C, Winsborrow M, Heyman J, Hall A (2017) Deglaciation of the Eurasian ice sheet complex. Quat Sci Rev 169:148–172

Iversen J (1934) Fund af Vildhest (*Equus caballus*) fra Overgangen mellem Sen- og Postglacialtid i Danmark. Meddelelser fra Dansk Geologisk Forening 8:327–340

Iversen J (1942) En pollenanalytisk Tidsfæstelse af Ferskvandlagene ved Nørre Lyngby. Meddelelser fra Dansk Geologisk Forening 10:130–151

Joo S, Han D, Lee EJ, Park S (2014) Use of length heterogeneity polymerase chain reaction (LH-PCR) as non-invasive approach for dietary analysis of Svalbard Reindeer, Rangifer tarandus platyrhynchus. PLoS One 9:e91552

Kater MM, Franken J, Carney KJ, Colombo L, Angenent GC (2001) Sex determination in the monoecious species cucumber is confined to specific floral whorls. Plant Cell 13:481–493

Kellmann-Sopyła W, Giełwanowska I (2015) Germination capacity of five polar Caryophyllaceae and Poaceae species under different temperature conditions. Polar Biol 38:1753–1765

Kellmann-Sopyła W, Giełwanowska I, Koc J, Górecki RJ, Domaciuk M (2017) Development of generative structures of polar Caryophyllaceae plants: the Arctic *Cerastium alpinum* and *Silene involucrata*, and the Antarctic *Colobanthus quitensis*. Polish Polar Res 38:83–104

Kim IS, Kim HY, Kim YS, Choi HG, Kang SH, Yoon HS (2013) Expression of dehydrin gene from arctic *Cerastium arcticum* increases abiotic stress tolerance and enhances the fermentation capacity of a genetically engineered *Saccharomyces cerevisiae* laboratory strain. Appl Microbiol Biotechnol 97:8997–9009

Kjær U, Olsen SL, Klanderud K (2018) Shift from facilitative to neutral interactions by the cushion plant *Silene acaulis* along a primary succession gradient. J Veg Sci 29:42–51

Kjellberg B, Karlsson S, Kerstensson I (1982) Effects of heliotropic movements of flowers of Dryas octopetala L. on gynoecium temperature and seed development. Oecologia 54:10–13

Kumar M, van Elsas JD, Nissinen R (2017a) Strong regionality and dominance of anaerobic bacterial taxa characterize diazotrophic bacterial communities of the arcto-alpine plant species *Oxyria digyna* and *Saxifraga oppositifolia*. Front Microbiol 8:1972

Kumar M, Brader G, Sessitsch A, Mäki A, van Elsas JD, Nissinen R (2017b) Plants assemble species specific bacterial communities from common core taxa in three arcto-alpine climate zones. Front Microbiol 8:12

Kumar M, Brader G, Sessitsch A, Mäki A, van Elsas JD, Nissinen R (2017c) Plants assemble species specific bacterial communities from common core taxa in three Arcto-Alpine climate zones. Front Microbiol 8:12

Kumar M, van Elsas JD, Nissinen R (2017d) Strong regionality and dominance of anaerobic bacterial taxa characterize diazotrophic bacterial communities of the Arcto-Alpine Plant Species *Oxyria digyna* and *Saxifraga oppositifolia*. Front Microbiol 8:1972

Kume A, Nakatsubo T, Bekku Y, Masuzawa T (1999) Ecological significance of different growth forms of purple saxifrage, *Saxifraga oppositifolia* L., in the high Arctic, Ny-Alesund, Norway. Arct Antarct Alp Res 31:27–33

La Farge C, Williams KH, England JH (2013) Regeneration of Little Ice Age bryophytes emerging from a polar glacier with implications of totipotency in extreme environments. Proc Natl Acad Sci U S A 110:9839–9844

Law R, Cook RED, Manlove RJ (1983) The ecology of flower and bulbil production in *Polygonum viviparum*. Nord J Bot 3:559–565

Levin DA, Soltis DE (2018) Factors promoting polyploid persistence and diversification and limiting diploid speciation during the K–Pg interlude. Curr Opin Plant Biol 42:1–7

Lin C, Thomashow MF (1992) DNA sequence analysis of a complementary DNA for cold-regulated *Arabidopsis* Gene cor15 and characterization of the COR 15 polypeptide. Plant Physiol 99:519–525

Lohaus R, Van de Peer Y (2016) Of dups and dinos: evolution at the K/Pg boundary. Curr Opin Plant Biol 30:62–69

Ma H (2005) Molecular genetic analyses of microsporogenesis and microgametogenesis in flowering plants. Annu Rev Plant Biol 56:393–434

Marr KL, Allen GA, Hebda RJ, McCormick LJ (2013) Phylogeographical patterns in the wide-spread arctic-alpine plant *Bistorta vivipara* (Polygonaceae) with emphasis on western North America. J Biogeogr 40:847–856

McGraw JB, Turner JB, Chandler JL, Vavrek MC (2014) Disturbances as hot spots of ecotypic variation: a case study with *Dryas octopetala*. Arct Antarct Alp Res 46:542–547

Mishler BD, Churchill SP (1984) A cladistic approach to the phylogeny of the "Bryophytes". Brittonia 36:406–424

Mishler BD, Lewis LA, Buchheim MA, Renzaglia KS, Garbary DJ, Delwiche CF, Zechman FW, Kantz TS Chapman RL (1994) Phylogenetic relationships of the "Green Algae" and "Bryophytes". Ann Mo Bot Gard 81:451–483

Molenda O, Reid A, Lortie CJ (2012) The alpine cushion plant Silene acaulis as foundation species: a bug's-eye view to facilitation and microclimate. PLoS One 7:e37223

Mooney HA, Billings WD (1961) Comparative physiological ecology of arctic and alpine populations of *Oxyria digyna*. Ecol Monogr 31:1–29

Morris W, Doak D (1998) Life history of the long-lived gynodioecious cushion plant *Silene acaulis* (Caryophyllaceae), inferred from size-based population projection matrices. Am J Bot 85:784–793

Müller E, Cooper EJ, Alsos IG (2011) Germinability of arctic plants is high in perceived optimal conditions but low in the field. Botany 89:337–348

Müller E, Eidesen PB, Ehrich D, Alsos IG (2012) Frequency of local, regional, and long-distance dispersal of diploid and tetraploid *Saxifraga oppositifolia* (Saxifragaceae) to Arctic glacier forelands. Am J Bot 99:459–471

Nakatsubo T, Fujiyoshi M, Yoshitake S, Koizumi H, Uchida M (2010) Colonization of the polar willow *Salix polaris* on the early stage of succession after glacier retreat in the High Arctic, Ny-Ålesund, Svalbard. Polar Res 29:285–390

Nansen F (1911) In Northern Mists: Arctic exploration in early times (complete). Library of Alexandria, pp 515–518

Neilson AH (1968) Vascular plants from the northern part of Nordaustlandet, Svalbard. Norsk Polarinstitutt Skrifter Nr. 143. Norsk Polaristitutt, Oslo

Neilson AH (1970) Vascular plants of Edgeøya, Svalbard. Norsk Polarinstitutt Skrifter Nr. 150. Norsk Polaristitutt, Oslo

Nickrent DL, Parkinson CL, Palmer JD, Duff RJ (2000) Multigene phylogeny of land plants with special reference to bryophytes and the earliest land plants. Mol Biol Evol 17:1885–1895

Nishiyama T, Wolf PG, Kugita M, Sinclair RB, Sugita M, Sugiura C, Wakasugi T, Yamada K, Yoshinaga K, Yamaguchi K, Ueda K, Hasebe M (2004) Chloroplast phylogeny indicates that bryophytes are monophyletic. Mol Biol Evol 21:1813–1819

Nissinen RM, Männistö MK, van Elsas JD (2012) Endophytic bacterial communities in three arctic plants from low arctic fell tundra are cold-adapted and host-plant specific. FEMS Microbiol Ecol 82:510–522

Norwegian Polar Institute. Place names of Svalbard and Jan Mayen. Last updated on 26 February 2010. Retrieved 10 July 2018

NSIU (1942) The place-names of Svalbard. Oslo

Oddveig Øien Ørvoll. Norwegian place names in polar regions. Last updated on 24 October 2016. Retrieved 10 July 2018

Øvstedal DO, Tønsberg T, Elvebakk A (2009) The lichen flora of Svalbard. Sommerfeltia 33:1–393

Pelaz S, Ditta GS, Baumann E, Wisman E, Yanofsky MF (2000) B and C floral organ identity functions require SEPALLATA MADS-box genes. Nature 405(6783):200–203

Pellino M, Hojsgaard D, Schmutzer T, Scholz U, Horand E, Vogel H, Sharbel TF (2013) Asexual genome evolution in the apomictic *Ranunculus* auricomus complex: examining the effects of hybridization and mutation accumulation. Mol Ecol 22:5908–5921

Platt DE, Haber M, Dagher-Kharrat MB, Douaihy B, Khazen G, Ashrafian Bonab M, Salloum A, Mouzaya F, Luiselli D, Tyler-Smith C, Renfrew C, Matisoo-Smith E, Zalloua PA (2017) Mapping post-glacial expansions: the peopling of Southwest Asia. Sci Rep 7:40338

Pucholt P, Hallingbäck HR, Berlin S (2017a) Allelic incompatibility can explain female biased sex ratios in dioecious plants. BMC Genom 18:251

Pucholt P, Wright AE, Conze LL, Mank JE, Berlin S (2017b) Recent sex chromosome divergence despite ancient dioecy in the willow *Salix viminalis*. Mol Biol Evol 34:1991–2001

Reid AM, Lortie CJ (2012) Cushion plants are foundation species with positive effects extending to higher trophic levels. Ecosphere 3:1–18

Rejment-Grochowska I (1967) Contribution to the Hepatic flora of the north coast of Hornsund (S. W. Svalbard). Acta Societatis Botanicorum Polloniae 36:531–544

Riechmann JL, Krizek AB, Meyerowitz EM (1996) Dimerization specificity of Arabidopsis MADS domain homeotic proteins APETALA1, APETALA3, PISTILLATA, and AGAMOUS. Proc Natl Acad Sci U S A 93(10):4793–4798

Riley AC, Ashlock DA, Graether SP (2019) Evolution of the modular, disordered stress proteins known as dehydrins. PLoS One 14:e0211813

Rønning O (1996) The flora of Svalbard. Norwegian Polar Institute, Oslo. 184 p

Ronse De Craene LP, Akeroyd JR (1988) Generic limits in Polygonum and related genera (Polygonaceae) on the basis of floral characters. Bot J Linn Soc 98:321–371

Ruhfel BR, Gitzendanner MA, Soltis PS, Soltis DE, Burleigh JG (2014) From algae to angiosperms-inferring the phylogeny of green plants (Viridiplantae) from 360 plastid genomes. BMC Evol Biol 14:23

Schönswetter P, Popp M, Brochmann C (2006) Rare arctic-alpine plants of the European Alps have different immigration histories: the snow bed species *Minuartia biflora* and *Ranunculus pygmaeus*. Mol Ecol 15:709–720

Schuster TM, Reveal JL, Kron KA (2011) Phylogeny of Polygoneae (Polygonaceae: Polygonoideae). Taxon 60:1653–1666

Schwarz-Sommer Z, Huijser P, Nacken W, Saedler H, Sommer H (1990) Genetic control of flower development by homeotic genes in *Antirrhinum majus*. Science 250(4983):931–936

Somerville C, Somerville S (1999) Plant functional genomics. Science 285(5426):380–383

Sommerfelt C (1832) Bidrag til Spetsbergens og Beeren-Eilands flora efter herbarier, medbragt af M. Keilhau. (Contributions to the flora of Spitsbergen and Bear Island based on herbaria brought by M. Keilhau.). Magazin for Naturvidenskaberne 11:232–252

Stenström M, Molau U (1992) Reproductive ecology of *Saxifraga oppositifolia*: phenology, mating system and reproductive success. Arct Alp Res 24:337–343

svalbardflora.net. Flora of Svalbard. Retrieved 10 July 2018

Svoen ME, Müller E, Brysting AK, Kålås IH, Eidesen PB (2019) Female advantage? Investigating female frequency and establishment performance in high-Arctic *Silene acaulis*. Botany 97:245–261

Teeri JA (1973) Polar desert adaptations of a high Arctic plant species. Science 179(4072):496–497

The Plant List (2013) Version 1.1. Published on the Internet; http://www.theplantlist.org/. Retrieved 1 Jan 2019

Tomita M, Masuzawa T (2010) Reproductive mode of depends on environment. Polar Sci 4:62–70

Unander S, Mortensen A, Elvebakk A (1985) Seasonal changes in crop content of the Svalbard Ptarmigan *Lagopus mutus hyperboreus*. Polar Res 3:239–245

Vik U, Carlsen T, Eidesen PB, Brysting AK, Kauserud H (2012) Microsatellite markers for *Bistorta vivipara* (Polygonaceae). Am J Bot 99:226–229

Walker DA, Breen AL, Raynolds MK, Walker MD (eds) (2013) Arctic Vegetation Archive (AVA) Workshop. CAFF Proceedings Report #10. Akureyri, Iceland

Wang Q, Liu J, Allen GA, Ma Y, Yue W, Marr KL, Abbott RJ (2016) Arctic plant origins and early formation of circumarctic distributions: a case study of the mountain sorrel, *Oxyria digyna*. New Phytol 209:343–353

Weretilnyk E, Orr W, White TC, Iu B, Singh J (1993) Characterization of three related low-temperature-regulated cDNAs from *Brassica napus*. Plant Physiol 101:171–177

Wikipedia. Svalbard. Last edited on 22 June 2018. Retrieved 10 July 2018

Wilhelm KS, Thomashow MF (1993) *Arabidopsis thaliana cor15b*, an apparent homologue of *cor15a*, is strongly responsive to cold and ABA, but not drought. Plant Mol Biol 23:1073–1077

Winkler M, Tribsch A, Schneeweiss GM, Brodbeck S, Gugerli F, Holderegger R, Abbott RJ, Schönswetter P (2012) Tales of the unexpected: phylogeography of the Arctic-alpine model plant *Saxifraga oppositifolia* (Saxifragaceae) revisited. Mol Ecol 21:4618–4630

Zawierucha K, Węgrzyn M, Ostrowska M, Wietrzyk P (2017) Tardigrada in Svalbard lichens: diversity, densities and habitat heterogeneity. Polar Biol 40:1385–1392

Zhou D, Zhou J, Meng L, Wang Q, Xie H, Guan Y, Ma Z, Zhong Y, Chen F, Liu J (2009) Duplication and adaptive evolution of the COR15 genes within the highly cold-tolerant *Draba* lineage (Brassicaceae). Gene 441:36–44

Chapter 5
The Past Shows the Future

Past: After the Ice Age

How can we know what kind of plants lived in the past in the Arctic? To understand the present, we need to know the past. In particular, to understand the current distribution of plants in the Arctic, you need to know the effects of the nearest past ice age on plants. What was the surviving plant at the edge of ice during the Ice Age? How did they expand their habitats in the Arctic after the last ice age? What was the plant that came to the Arctic as the Ice Age retreated and the ice disappeared?

But knowing the past is not easy. It would be nice to have a lot of fossils, but plant fossils of the Ice Age are not common. Scientists who realized that they could not rely on fossils have observed pollen remaining in a lake or coastal sediments, or parts of plants that are not fully decomposed in the ground. However, this was time-consuming work, and it was not easy to identify plant species with pollen or leaf debris.

DNA analysis, ancient DNA metabarcoding has opened a breakthrough to identify plants. DNA extraction from soils or sediments was already developed. Soils and sediments contain a variety of phenolic compounds, it was much harder than extracting DNA from living organisms. As the kit for extracting DNA from soil emerged, scientists could access the DNA that was left in nature. The problem is that the old DNA breaks into small pieces over time. Because the size of genes used to analyze the living plant exceeds hundreds of base pairs, small fragments cannot be used to identify the plant. New DNA analysis tools were needed to obtain information for identifying plants in DNA fragments.

Pierre Taberlet at Alpine Ecology Laboratory in Université Grenoble Alpes, France, and his research colleagues had found that a short part of a gene encoding chloroplast transfer RNA for leucine, trnL (UAA) is suitable to be used for fragmented DNA. If two regions of an RNA strand have complementary nucleotide sequences, they make hydrogen bonds and partially a double helix that ends in an unpaired loop. The double helix is conserved because mutation does not occur

© Springer Nature Switzerland AG 2020
Y. K. Lee, *Arctic Plants of Svalbard*, https://doi.org/10.1007/978-3-030-34560-0_5

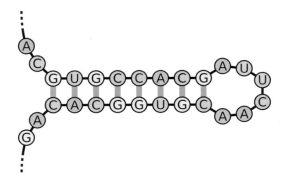

easily, while the loop region is less conservative. This is called a "hairpin loop"
structure because its shape resembles a hairpin (Fig. 5.1). DNA sequences with
hairpin loop structures are appropriately variable enough to distinguish the organisms from each other and are conserved enough to be compared with each other. The
P6 loop in the trnL (UAA) intron is the DNA sequences with hairpin loop structures.
It is very short fragments (10–143 bp) enough to be amplified from ancient DNA.
Using next-generation sequencing, metagenomic analysis can be performed with
DNA extracted from old layers of sediments. Using P6 loop in the trnL (UAA)
intron, 100% of families and ca 90% of genera of Arctic plants was identified,
whereas less than 50% of species could be identified.

Plant diversity has declined during the Last Glacial Maximum according to the
analysis using P6 loop metabarcoding of samples from Siberia, Canadian High
Arctic, Alaska, and Svalbard. About one-third of the plants that had inhabited the
Arctic before the Ice Age survived to the present through the Ice Age. A quarter of
the plants that had inhabited the Arctic before the Ice Age have disappeared during
the Ice Age and are now living in the Arctic again. It was known that Arctic
mammoth steppe during the Late Quaternary was dominated by graminoid. But
the research group studying P6 loop metabarcoding of ancient DNA insisted that the
Arctic vegetation had been dry steppe-tundra dominated by forbs not by graminoid.
The Arctic vegetation includes *Bromus pumpellianus* and *Puccinellia* sp. (Poaceae),
Artemisia frigida (Asteraceae), *Plantago canescens* (Plantaginaceae), *Anemone
patens* and *Caltha* sp. (Ranunculaceae), *Armeria* sp. (Plumbaginaceae), *Dryas*
sp. (Rosaceae), *Draba* sp. (Brassicaceae), and *Salix* sp. (Salicaceae). *Equisetum*
(Equisetaceae) and *Eriophorum* (Cyperaceae) were also detected in the P6 loop
metabarcoding, which means that there were moist conditions (Fig. 5.2). The data
showed that after the Ice Age, dry steppe-tundra had changed into moist tundra.

In the Ice Age, Svalbard was almost completely covered with ice. Therefore,
Svalbard plants are likely to have migrated from elsewhere after the Ice Age.
Population genetics gave us valuable insight on plant migration. The data of *Dryas
octopetala, Salix herbacea, Cassiope tetragona, Saxifraga rivularis*, and five more
species showed at least 6~38 plants per each species must have successfully moved
to and survived in Svalbard from northwestern Russia, eastern Greenland, and

Fig. 5.2 Svalbard plants of *Equisetum* (left) and *Eriophorum* (right)

Scandinavia. The vectors of the plant migration are wind, sea ice, drift wood, birds, and mammals. In particular, sea ice makes concrete ways between northwestern Russia and Svalbard during winter. Eastern Greenland is also frequently connected to Svalbard. Migratory birds from Scandinavia, Iceland, and Scotland visit Svalbard every summer. Through these different vectors, Arctic plants living outside Svalbard became an insider.

This long-distance dispersal of Arctic plants has happened a long time ago in the past. Based on this past and temperature increase in Svalbard, we can predict that more Arctic plants will be able to migrate to Svalbard in the near future.

Future: Climate Change and the Arctic Plants

Nature is not simple as it seems. Various environmental factors work together in nature. Climate change causes not only temperature rise but also the complex changes of various environmental factors. Climate change may be a crisis for some Arctic tundra plants. During the past 30 years, temperatures in the Arctic tundra have increased by around 1 °C in summer and 1.5 °C in winter. Scientists say, "The Arctic is warming up faster than almost anywhere else on earth."

To predict the future, we need to know the changes in the recent past. Several decadal monitoring has confirmed that the increase of air temperature is a reality at the Arctic. So, what changes will temperature increase bring to Arctic vegetation? Over three decades of warming at 117 tundra locations showed that summer temperature is positively associated with plant community height and specific leaf area. The plant community height has increased over time, due to the movement of tall species upward from nearby warmer microclimates. One research paper said, "If taller plants continue to spread at the current rate, the plant community height could increase by 20% to 60% by the end of the century." Another study showed that height increases occurred in deciduous shrubs, tall shrubs, graminoids (especially grasses) and forbs, whereas mosses decreased. The increase of plant community height can have feedback effects such as woody litter production and carbon storage increase, soil temperature decreases by shading, albedo reduction, snow accumulation increase, and so on. Therefore, we have to consider these various feedbacks to predict the consequence of climate change in the Arctic tundra.

The temperature effect can be enhanced by soil moisture. The relationship between temperature and specific leaf area was much stronger where soil moisture was high. The leaf dry matter content decreased according to temperature increase in wet sites, but no change in dry sites. Long-term climate warming should cause the shift to higher communities with taller plants, larger leaf area, and less leaf dry matter content at wetter sites. Temperature and moisture are strongly related, and the soil temperature increase can also lead to drier soils in the long term. To understand the shifts in tundra plant communities by future warming, the changes in water availability should be considered.

Climate change is causing a variety of changes in Arctic vegetation. The contemporary warming in the Arctic has already resulted in the increase of plant productivity. Some data showed the increases in abundance of evergreen and low-growing tall shrubs, and the decrease of bare ground. Although shrubs increased, litter accumulation did not increase. It is possible that the warming promoted litter decomposition by microbes, and net accumulation did not change. Summer temperature increase did not seem to reduce lichens and mosses. Their decrease is thought to be an indirect result of warming such as shading. Long-term summer warming is closely related to the changes in vascular plant composition. The patterns of change vary from region to region.

The response to climate change also varies depending on the plant species. The climate change manipulation experiments did not improve the growth and reproduction of some Arctic tundra plants. For example, warming had no significant effect on *Ranunculus glacialis*. Date of snowmelt is associated with growth and flowering of Arctic plants. Mild weather during the snow-free period can greatly impact the Arctic vegetation. Although the average date of snowmelt increased due to warming, it had no impact on seed number and weight, or leaf number and width, rather, the seed set decreased. *R. glacialis* seems to adapt to low temperature. On the other hand, the seed set of *R. acris* increased significantly when temperature increased. In conclusion, *R. glacialis* does not seem to be sensitive to temperature change. When other species with higher sensitivity and phenotypic plasticity to temperature change invade in Arctic tundra, *R. glacialis* is likely to reduce in the habitat. *R. glacialis* is a vulnerable species on the Red List of Svalbard plants. There might be competition with other plants that have expanded their habitat to the Arctic, which is disadvantageous for arctic tundra plants such as *R. glacialis*.

Plants are not only affected by climate, herbivores can alter the Arctic plant communities. Soil microbiome that has been overlooked so far may affect plant communities, and soil pH is the key to determine soil microbiome. Precipitation and permafrost stability are also important factors to determine Arctic vegetation, which may affect soil moisture. Snow cover and ice on snow are neglected drivers, and they can cause winter disasters. The shrub itself may cause a change. Increased shrubs accelerate snowmelt and sensible heat flux during snowmelt, and they can cause cooler summer and warmer winter soils. Shrubs also affect abundance and diversity of moss and lichen. The plant community and the environment are closely connected to each other directly or indirectly, leading to reaction feedback when one part of the ecosystem changes.

Scientists say that the Arctic "ecosystems heat up, tundra changes speed up."

Bibliography

Alsos IG, Eidesen PB, Ehrich D, Skrede I, Westergaard K, Jacobsen GH, Landvik JY, Taberlet P, Brochmann C (2007) Frequent longdistance colonization in the changing Arctic. Science 316:1606–1609

Alsos IG, Sjögren P, Edwards ME, Landvik JY, Gielly L, Forwick M, Coissac E, Brown AG, Jakobsen LV, Føreid MK (2016) Sedimentary ancient DNA from Lake Skartjørna, Svalbard: assessing the resilience of arctic flora to Holocene climate change. The Holocene 26:627–642

Bjorkman AD, Myers-Smith IH, Elmendorf SC et al (2018a) Tundra Trait Team: a database of plant traits spanning the tundra biome. Glob Ecol Biogeogr 27:1402–1411

Bjorkman AD, Myers-Smith IH, Elmendorf SC et al (2018b) Plant functional trait change across a warming tundra biome. Nature 562(7725):57–62

Elmendorf SC, Henry GH, Hollister RD et al (2012a) Global assessment of experimental climate warming on tundra vegetation: heterogeneity over space and time. Ecol Lett 15:164–175

Elmendorf SC, Henry GHR, Hollister RD et al (2012b) Plot-scale evidence of tundra vegetation change and links to recent summer warming. Nat Clim Chang 2:453–457

Ernakovich JG, Hopping KA, Berdanier AB, Simpson RT, Kachergis EJ, Steltzer H, Wallenstein MD (2014) Predicted responses of arctic and alpine ecosystems to altered seasonality under climate change. Glob Chang Biol 20:3256–3269

https://www.pnnl.gov/science/highlights/highlight.asp?id=5012

Myers-Smith IH, Elmendorf SC, Beck PSA et al (2015) Climate sensitivity of shrub growth across the tundra biome. Nat Clim Chang 5:887–891

Myers-Smith IH, Thomas HJD, Bjorkman AD (2019) Plant traits inform predictions of tundra responses to global change. New Phytol 221:1742–1748

Niittynen P, Heikkinen RK, Luoto M (2018) Snow cover is a neglected driver of Arctic biodiversity loss. Nat Clim Chang 8:997–1001

Prevéy J, Vellend M, Rüger N et al (2017) Greater temperature sensitivity of plant phenology at colder sites: implications for convergence across northern latitudes. Glob Chang Biol 23:2660–2671

Skrede I, Eidesen PB, Portela RP, Brochmann C (2006) Refugia, differentiation and postglacial migration in arctic-alpine Eurasia, exemplified by the mountain avens (Dryas octopetala L.). Mol Ecol 15:1827–1840

Sønstebø JH, Gielly L, Brysting AK, Elven R, Edwards M, Haile J, Willerslev E, Coissac E, Rioux D, Sannier J, Taberlet P, Brochmann C (2010) Using next-generation sequencing for molecular reconstruction of past Arctic vegetation and climate. Mol Ecol Resour 10:1009–1018

Taberlet P, Coissac E, Pompanon F, Gielly L, Miquel C, Valentini A, Vermat T, Corthier G, Brochmann C, Willerslev E (2007) Power and limitations of the chloroplast trnL (UAA) intron for plant DNA barcoding. Nucleic Acids Res 35(3):e14

Thomas HJD, Myers-Smith IH, Bjorkman AD et al (2019) Traditional plant functional groups explain variation in economic but not size-related traits across the tundra biome. Glob Ecol Biogeogr 28:78–95

Totland Ø, Alatalo JM (2002) Effects of temperature and date of snowmelt on growth, reproduction, and flowering phenology in the arctic/alpine herb, Ranunculus glacialis. Oecologia 133:168–175

Willerslev E, Davison J, Moora M, Zobel M, Coissac E, Edwards ME, Lorenzen ED, Vestergård M, Gussarova G, Haile J (2014) Fifty thousand years of Arctic vegetation and megafaunal diet. Nature 506:47–51
Yoccoz NG, Bråthen KA, Gielly L, Haile J, Edwards ME, Goslar T, Von Stedingk H, Brysting AK, Coissac E, Pompanon F, Sønstebø JH, Miquel C, Valentini A, De Bello F, Chave J, Thuiller W, Wincker P, Cruaud C, Gavory F, Rasmussen M, Gilbert MT, Orlando L, Brochmann C, Willerslev E, Taberlet P (2012) DNA from soil mirrors plant taxonomic and growth form diversity. Mol Ecol 21:3647–3655

List of Svalbard Vascular Plants

Scientific name	English common name
Lycopodiophyta	
Lycopodiopsida	
Lycopodiales	
Lycopodiaceae	
Huperzia arctica[a]	Mountain Fir-moss
Lycopodium selago[b]	Mountain Club-moss
Pteridophyta	
Equisetopsida	
Equisetales	
Equisetaceae	
Equisetum arvense ssp. *alpestre*[c]	Polar Horsetail
Equisetum scirpoides	Dwarf Horsetail
Equisetum variegatum ssp. *variegatum*[d]	Variegated Horsetail
Psilotopsida	
Ophioglossales	
Ophioglossaceae	
Botrychium boreale	Northern Moonwort
Botrychium lunaria	Moonwort
Polypodiopsida	
Polypodiales	
Cystopteridaceae	
Cystopteris fragilis[e]	Brittle Bladder-fern
Woodsiaceae	
Woodsia glabella	Smooth Woodsia
Magnoliophyta	
Alismatales	
Tofieldiaceae	
Tofieldia pusilla	Scottish Asphodel
Poales	
Cyperaceae	

(continued)

© Springer Nature Switzerland AG 2020
Y. K. Lee, *Arctic Plants of Svalbard*, https://doi.org/10.1007/978-3-030-34560-0

Scientific name	English common name
Carex bigelowii ssp. *arctisibirica*	Bigelow's Sedge
Carex capillaris ssp. *fuscidula*[f]	Hair Sedge
Carex concolor[g]	Water Sedge
Carex fuliginosa ssp. *misandra*[h]	Shortleaved Sedge
Carex glacialis	Glacial Sedge
Carex glareosa	Gravel Sedge
Carex krausei	Krause's Sedge
Carex lachenalii	Twotipped Sedge
Carex lidii	Lids Sedge
Carex marina ssp. *pseudolagopina*[i]	Sea Sedge
Carex maritima	Curved Sedge
Carex nardina ssp. *hepburnii*[j]	Hepburn's Sedge
Carex parallela	Parallel Sedge
Carex rupestris	Curly Sedge
Carex saxatilis ssp. *laxa*[k]	Rock Sedge
Carex subspathacea	Hoppner's Sedge
Carex ursina	Bear Sedge
Eriophorum scheuchzeri ssp. *arcticum*[l]	Arctic Cottongrass
Eriophorum sorensenii	Rousseau's Cottongrass
Eriophorum triste[m]	Narrowleaf Cottongrass
Kobresia simpliciuscula ssp. *subholarctica*[n]	Simple Bog Sedge
Poales **Juncaceae**	
Juncus albescens[o]	Northern White Rush
Juncus arcticus	Arctic Rush
Juncus biglumis	Twoflower Rush
Juncus leucochlamys[p]	Chestnut Rush
Luzula arcuata[q]	Curved Woodrush
Luzula confusa[r]	Northern Woodrush
Luzula nivalis[s]	Reindeer Woodrush
Luzula wahlenbergii	Arctic Woodrush
Poales **Poaceae**	
Agrostis capillaris[t]	Colonial Bentgrass
Alopecurus ovatus[u]	Polar Foxtail
Arctagrostis latifolia	Russian Grass
Arctodupontia scleroclada	Kongsfjord Grass
Arctophila fulva	Arctic Marsh Grass
Calamagrostis neglecta ssp. *groenlandica*[v]	Narrow Small-reed
Calamagrostis purpurascens	Purple Reedgrass
Deschampsia alpina[w]	Alpine Hairgrass
Deschampsia cespitosa	Tufted Hairgrass
Deschampsia sukatschewii ssp. *borealis*[x]	Tundra Hairgrass
Dupontia fisheri[y]	Fisher's Tundragrass

(continued)

Scientific name	English common name
Festuca baffinensis	Baffin Fescue
Festuca brachyphylla	Alpine Fescue
Festuca edlundiae	Grass-cushion
Festuca hyperborea	Boreal Fescue
Festuca rubra ssp. *richardsonii*[z]	Richardson's Fescue
Festuca vivipara	Viviparous Fescue
Festuca viviparoidea	Northern Fescue
Hierochloe alpina	Alpine Sweetgrass
Pleuropogon sabinei[aa]	False Semaphoregrass
Poa abbreviata	Short Bluegrass
Poa alpigena[ab]	Smooth Meadow-grass
Poa alpina var. *vivipara*[ac]	Alpine Meadow-grass
Poa arctica[ad]	Arctic Bluegrass
Poa glauca[ae]	Glaucous Meadow-grass
Poa hartzii[af]	Hartz's Bluegrass
Poa pratensis[ag]	Smooth Meadow-grass
Puccinellia angustata	Northern Alkaligrass
Puccinellia coarctata[ah]	Sea Urchins Alkaligrass
Puccinellia phryganodes	Creeping Alkaligrass
Puccinellia svalbardensis[ai]	Svalbard Alkaligrass
Puccinellia vahliana[aj]	Vahl's Alkaligrass
Pucciphippsia vacillans[ak]	Svalbard Ggrass
Trisetum spicatum	Spike Trisetum
Ranunculales **Papaveraceae**	
Papaver cornwallisense[al]	
Papaver dahlianum[am]	Svalbard Poppy
Ranunculales **Ranunculaceae**	
Coptidium lapponicum[an]	Lapland Buttercup
Coptidium pallasii[ao]	Glossy Buttercup
Ranunculus acris[ap]	Arctic Buttercup
Ranunculus arcticus	Tall Buttercup
Ranunculus glacialis[aq]	Glacier Buttercup
Ranunculus hyperboreus ssp. *arnellii*	Tundra Buttercup
Ranunculus hyperboreus ssp. *hyperboreus*[ar]	Swamp Buttercup
Ranunculus nivalis	Snow Buttercup
Ranunculus pedatifidus[as]	Birdfoot Buttercup
Ranunculus pygmaeus	Pigmy Buttercup
Ranunculus repens	Creeping Buttercup
Ranunculus subborealis ssp. *villosus*	Meadow Buttercup
Ranunculus sulphureus	Sulphur Buttercup
Ranunculus wilanderi	Polar Kidney Buttercup

(continued)

Scientific name	English common name
Caryophyllales	
Caryophyllaceae	
Arenaria humifusa	Low Sandworts
Arenaria pseudofrigida[at]	Fringed Sandworts
Cerastium alpinum	Alpine Chickweed
Cerastium arcticum	Arctic Mouse-ear
Cerastium cerastoides	Mountain Chickweed
Cerastium regelii ssp. *caespitosum*[au]	Regel's Chickweed
Honckenya peploides ssp. *diffusa*[av]	Sea Sandwort
Minuartia biflora	Mountain Sandwort
Minuartia rossii	Ross' Sandwort
Minuartia rubella	Beautiful Sandwort
Minuartia stricta	Rock Sandwort
Sagina caespitosa	Tufted Pearlwort
Sagina nivalis[aw]	Snow Pearlwort
Silene acaulis	Moss Campion
Silene involucrata ssp. *furcata*[ax]	Artic White Campion
Silene uralensis ssp. *arctica*[ay]	Polar Campion
Silene uralensis ssp. *apetala*[az]	Apetalous Catchfly
Stellaria humifusa	Saltmarsh Starwort
Stellaria longipes	Longstalk Starwort
Stellaria media	Chickweed
Caryophyllales	
Polygonaceae	
Bistorta vivipara[ba]	Alpine Bistort
Koenigia islandica	Iceland Purslane
Oxyria digyna	Mountain Sorrel
Rumex acetosa	Common Sorrel
Rumex acetosella[bb]	Sheep's Sorrel
Saxifragales	
Crassulaceae	
Rhodiola rosea[bc]	Roseroot
Saxifragales	
Saxifragaceae	
Chrysosplenium tetrandrum	Northern Golden Saxifrage
Micranthes foliolosa[bd]	Foliolose Saxifrage
Micranthes hieraciifolia[be]	Stiff Stem Saxifrage
Micranthes nivalis[bf]	Snow Saxifrage
Micranthes tenuis[bg]	Ottertail Pass Saxifrage
Saxifraga aizoides	Yellow Saxifrage
Saxifraga cernua	Drooping Saxifrage
Saxifraga cespitosa	Tufted Saxifrage
Saxifraga hirculus ssp. *compacta*[bh]	Marsh Saxifrage
Saxifraga hyperborea	Pygmy Saxifrage

(continued)

Scientific name	English common name
Saxifraga oppositifolia	Purple Saxifrage
Saxifraga platysepala[bi]	Thread Saxifraga
Saxifraga rivularis	Alpine Brook Saxifrage
Saxifraga svalbardensis	Svalbard Saxifrage
Brassicales **Brassicaceae**	
Arabis alpina	Alpine Rock-cress
Barbarea vulgaris	Winter-cress
Braya glabella ssp. *purpurascens*[bj]	Purplish Braya
Cakile maritima ssp. *islandica*[bk]	Arctic Sea Rocket
Capsella bursa-pastoris[bl]	Shephered's Purse
Cardamine bellidifolia ssp. *bellidifolia*[bm]	Alpine Cress
Cardamine pratensis ssp. *angustifolia*[bn]	Polar Cress
Cardamine silvestris[bo]	Forest Cress
Cochlearia groenlandica	Greenland Scurvygrass
Draba alpina	Alpine Whitlowgrass
Draba arctica[bp]	Arctic Draba
Draba cinerea[bq]	Gray-leaf Draba
Draba corymbosa[br]	Flat-top Draba
Draba daurica[bs]	Smooth Draba
Draba fladnizensis	Austrian Draba
Draba glabella	Scatter Draba
Draba lactea	Lapland Whitlowgrass
Draba micropetala	Small-flowered Draba
Draba nivalis	Snow Whitlowgrass
Draba norvegica[bt]	Rock Whitlowgrass
Draba oxycarpa[bu]	Gredin's Whitlowgrass
Draba pauciflora	Few-flowered Whitlowgrass
Draba subcapitata	Ellesmere Island Whitlowgrass
Eutrema edwardsii	Edwards' Mock Wallflower
Thlaspi arvense	Field Penny-cress
Malpighiales **Salicaceae**	
Salix arctica[bv]	Arctic Willow
Salix herbacea	Dwarf Willow
Salix lanata	Woolly Willow
Salix polaris	Polar Willow
Salix reticulata	Netleaf Willow
Fabales **Fabaceae**	
Trifolium repens	White Clover
Rosales **Rosaceae**	
Alchemilla glomerulans	Clustered Lady's Mantle

(continued)

Scientific name	English common name
Alchemilla subcrenata	Broadtooth Lady's Mantle
Alchemilla vulgaris[bw]	Lady's Mantle
Dryas octopetala	Mountain Avens
Potentilla arenosa ssp. *chamissonis*	Bluff Cinquefoil
Potentilla crantzii	Alpine Cinquefoil
Potentilla hyparctica[bx]	Arctic Cinquefoil
Potentilla insularis	Svalbard Cinquefoil
Potentilla lyngei[by]	Lynge Cinquefoil
Potentilla nivea[bz]	Snow Cinquefoil
Potentilla pulchella	Pretty Cinquefoil
Potentilla rubricaulis[ca]	Rocky Mountain Cinquefoil
Rubus chamaemorus	Cloudberry
Sibbaldia procumbens	Creeping Sibbaldia
Fagales **Betulaceae**	
Betula nana ssp. *tundrarum*[cb]	Dwarf Birch
Ericales **Ericaceae**	
Cassiope tetragona	Arctic Bell-heather
Empetrum nigrum ssp. *hermaphroditum*[cc]	Mountain Crowberry
Harrimanella hypnoides	Moss Bell-heather
Vaccinium uliginosum ssp. *microphyllum*[cd]	Polar Bilberry
Ericales **Polemoniaceae**	
Polemonium boreale	Northern Jacob's Ladder
Boraginales **Boraginaceae**	
Mertensia maritima ssp. *tenella*[ce]	Baltic Wort
Gentianales **Gentianaceae**	
Comastoma tenellum[cf]	Slender Gentian
Lamiales **Lentibulariaceae**	
Pinguicula alpina	Alpine Butterwort
Lamiales **Orobanchaceae**	
Pedicularis dasyantha	Wooly Lousewort
Pedicularis hirsuta	Hairy Lousewort
Lamiales **Plantaginaceae**	
Hippuris lanceolata	Lance-leaved Mare's-tail
Hippuris vulgaris[cg]	Mare's Tail
Lamiales **Scrophulariaceae**	
Euphrasia wettsteinii	Mountain Eyebright

(continued)

Scientific name	English common name
Asterales	
Asteraceae	
Achillea millefolium	Yarrow
Arnica angustifolia[ch]	Alpine Arnica
Erigeron eriocephalus[ci]	One Flower Fleabane
Erigeron humilis[cj]	Snow Fleabane
Petasites frigidus	Arctic Butterbur
Saussurea alpina	Alpine Saw-wort
Taraxacum acromaurum[ck]	Iceland Dandelion
Taraxacum arcticum	Arctic Dandelion
Taraxacum brachyceras	Common Dandelion
Asterales	
Campanulaceae	
Campanula rotundifolia ssp. *gieseckiana*[cl]	Hairbell
Campanula uniflora	Alpine Hairbell
Apiales	
Apiaceae	
Anthriscus sylvestris	Cow Parsley

[a] *Huperzia selago* ssp. *arctica* in Rønning (1996) and Alsos et al. (2004)
[b] It was listed not in The Flora of Svalbard (http://www.svalbardflora.net) but in the previous reports (Dahl 1937; Neilson 1968, 1970)
[c] *Equisetum arvense* in The Plant List (http://www.theplantlist.org); *E. arvense* ssp. *boreale* in Alsos et al. (2004) and Evju et al. (2010)
[d] *Equisetum variegatum* in The Plant List (http://www.theplantlist.org)
[e] *Cystopteris dickieana* in Rønning (1996)
[f] *Carex capillaris* in Rønning (1996)
[g] *Carex aquatilis* var. *minor* in The Plant List (http://www.theplantlist.org); *C. stans* in Rønning (1996); *C. aquatilis* ssp. *stans* in Lydersen (2010)
[h] *Carex fuliginosa* in The Plant List (http://www.theplantlist.org); *Carex misandra* in the previous reports (Dahl 1937; Neilson 1968, 1970; Rønning 1996)
[i] *Carex amblyorhyncha* in Rønning (1996)
[j] *Carex nardina* var. *hepburnii* in The Plant List (http://www.theplantlist.org)
[k] *Carex saxatilis* in The Plant List (http://www.theplantlist.org)
[l] *Eriophorum scheuchzeri* in The Plant List (http://www.theplantlist.org)
[m] *Eriophorum angustifolium* ssp. *triste* in The Plant List (http://www.theplantlist.org)
[n] *Kobresia simpliciuscula* in Rønning (1996)
[o] *Juncus triglumis* ssp. *albescens* in The Plant List (http://www.theplantlist.org)
[p] *Juncus castaneus* ssp. *leucochlamys* in The Plant List (http://www.theplantlist.org)
[q] It was listed not in The Flora of Svalbard (http://www.svalbardflora.net) but in Rønning (1996). *Luzula arcuata* ssp. *arcuata* in Engelskjon (1986) and Alsos et al. (2004)
[r] It was listed not in The Flora of Svalbard (http://www.svalbardflora.net) but in the previous reports (Dahl 1937; Neilson 1968, 1970; Rønning 1996; Evju et al. 2010). *Luzula arcuata* ssp. *confusa* in Alsos et al. (2004)
[s] *Luzula arctica* in the previous reports (Dahl 1937; Neilson 1968, 1970; Rønning 1996)
[t] It was listed not in The Flora of Svalbard (http://www.svalbardflora.net) but in Engelskjon (1986)
[u] *Alopecurus magellanicus* in The Plant List (http://www.theplantlist.org), and *A. borealis* in Rønning (1996) and Alsos et al. (2004)
[v] *Calamagrostis stricta* in The Plant List (http://www.theplantlist.org); *C. neglecta* in Engelskjon (1986)

(continued)

[w] *Deschampsia cespitosa* in The Plant List (http://www.theplantlist.org); *D. borealis* in Rønning (1996) and Alsos et al. (2004)

[x] *Deschampsia cespitosa* in The Plant List (http://www.theplantlist.org)

[y] *Dupontia fisheri* ssp. *psilosantha* in Dahl (1937); *D. pelligera* in Rønning (1996); *D. psilosantha* in the previous reports (Engelskjon 1986; Rønning 1996; Alsos et al. 2004; Evju et al. 2010)

[z] *Festuca richardsonii* in The Plant List (http://www.theplantlist.org); *F. cryophila* in Rønning (1996); *F. richardsonii* ssp. *cryophila* in Neilson (1970); *F. rubra* ssp. *arctica* in Alsos et al. (2004); *F. rubra* var. *arenaria* in Dahl (1937)

[aa] *Pleuropogon sabinii* in Rønning (1996)

[ab] It was listed not in The Flora of Svalbard (http://www.svalbardflora.net) but in the previous reports (Neilson 1968, 1970; Evju et al. 2010). Subspecies and variation were also reported as *Poa alpigena* var. *colpodea* in Dahl (1937) and Neilson (1970), and as *P. pratensis* ssp. *alpigena* in Engelskjon (1986)

[ac] It was listed not in The Flora of Svalbard (http://www.svalbardflora.net) but in Dahl (1937). *Poa alpina* in The Plant List (http://www.theplantlist.org); *P. alpina* var. *alpina* in Engelskjon (1986)

[ad] It was listed not in The Flora of Svalbard (http://www.svalbardflora.net). *Poa alpigena* var. *vivipara* in the previous reports (Dahl 1937; Neilson 1970; Evju et al. 2010); *Poa arctica* var. *vivipara* in Dahl (1937) and Neilson (1970)

[ae] It was listed not in The Flora of Svalbard (http://www.svalbardflora.net) but in the previous reports (Rønning 1996; Alsos et al. 2004; Evju et al. 2010)

[af] It was listed not in The Flora of Svalbard (http://www.svalbardflora.net) but in Rønning (1996)

[ag] It was listed not in The Flora of Svalbard (http://www.svalbardflora.net) but in Rønning (1996)

[ah] *Puccinellia distans* in The Plant List (http://www.theplantlist.org); *P. capillaris* in Rønning (1996)

[ai] *Puccinellia tenella* in The Plant List (http://www.theplantlist.org) and Rønning (1996)

[aj] *Colpodium vahlianum* in the previous reports (Dahl 1937; Neilson 1968, 1970; Rønning 1996)

[ak] *Colpodium vaeillans* in the previous reports (Dahl 1937; Neilson 1968, 1970; Rønning 1996)

[al] *Papaver radicatum* in The Plant List (http://www.theplantlist.org)

[am] *Papaver dahlianum* ssp. *polare* in Evju et al. (2010)

[an] *Ranunculus lapponicus* in Rønning (1996)

[ao] *Ranunculus pallasii* in Rønning (1996)

[ap] It was listed not in The Flora of Svalbard (http://www.svalbardflora.net) but in Rønning (1996)

[aq] *Beckwithia glacialis* in The Plant List (http://www.theplantlist.org)

[ar] *Ranunculus hyperboreus* in The Plant List (http://www.theplantlist.org)

[as] It was listed not in The Flora of Svalbard (http://www.svalbardflora.net) but in Neilson (1970). *Ranunculus affinis* in Rønning (1996)

[at] *Arenaria ciliata* ssp. *pseudofrigida* in Dahl (1937)

[au] *Cerastium regelii* in The Plant List (http://www.theplantlist.org)

[av] *Honckenya peploides* in The Plant List (http://www.theplantlist.org)

[aw] *Sagina intermedia* in Dahl (1937) and Neilson (1970)

[ax] *Silene involucrata* in The Plant List (http://www.theplantlist.org); *S. furcata* in Rønning (1996) and Alsos et al. (2004); *Melandrium affine* in Dahl (1937)

[ay] *Silene uralensis* in The Plant List (http://www.theplantlist.org); *Melandrium apetalum* ssp. *arcticum* in the previous reports (Dahl 1937; Neilson 1968, 1970)

[az] It was listed not in The Flora of Svalbard (http://www.svalbardflora.net) but in Neilson (1970). *Melandrium apetalum* in Dahl (1937)

[ba] *Persicaria vivipara* in The Plant List (http://www.theplantlist.org); *Polygonum viviparum* in the previous reports (Dahl 1937; Neilson 1968, 1970; Rønning 1996)

[bb] It was listed not in The Flora of Svalbard (http://www.svalbardflora.net) but in Rønning (1996)

[bc] *Rhodiola arctica* in Rønning (1996); *Sedum roseum* in The Plant List (http://www.theplantlist.org); *Sedum rosea* ssp. *arcticum* in Engelskjon (1986)

[bd] *Saxifraga foliolosa* in The Plant List (http://www.theplantlist.org)

[be] *Saxifraga hieracifolia* in The Plant List (http://www.theplantlist.org)

(continued)

[bf] *Saxifraga nivalis* in The Plant List (http://www.theplantlist.org)
[bg] *Saxifraga tenuis* in The Plant List (http://www.theplantlist.org)
[bh] *Saxifraga hirculus* in Rønning (1996); *S. hirculus* ssp. *alpina* in The Plant List (http://www.theplantlist.org)
[bi] *Saxifraga flagellaris* ssp. *platysepala* in The Plant List (http://www.theplantlist.org)
[bj] *Braya glabella* in The Plant List (http://www.theplantlist.org)
[bk] *Cakile arctica* in The Plant List (http://www.theplantlist.org)
[bl] It was listed not in The Flora of Svalbard (http://www.svalbardflora.net) but in Rønning (1996)
[bm] *Cardamine bellidifolia* in The Plant List (http://www.theplantlist.org)
[bn] *Cardamine pratensis* ssp. *polemonioides* in The Plant List (http://www.theplantlist.org); *C. nymanii* in the previous reports (Dahl 1937; Neilson 1968, 1970; Engelskjon 1986; Rønning 1996)
[bo] *Rorippa silvestris* in The Plant List (http://www.theplantlist.org) and in Korean Plant Names Index (http://www.nature.go.kr)
[bp] *Draba oblongata* in The Plant List (http://www.theplantlist.org); *D. arctica* ssp. *groenlandiea* in Neilson (1968)
[bq] It was listed not in The Flora of Svalbard (http://www.svalbardflora.net) but in Neilson (1970) and Evju et al. (2010)
[br] *Draba bellii* in the previous reports (Dahl 1937; Neilson 1968, 1970)
[bs] It was listed not in The Flora of Svalbard (http://www.svalbardflora.net) but in the previous reports (Neilson 1968; Rønning 1996; Alsos et al. 2004)
[bt] *Draba rupestris* in Dahl (1937)
[bu] *Draba gredinii* in the previous reports (Dahl 1937; Neilson 1968, 1970)
[bv] It was listed not in The Flora of Svalbard (http://www.svalbardflora.net) but in Rønning (1996)
[bw] It was listed not in The Flora of Svalbard (http://www.svalbardflora.net) but in Rønning (1996)
[bx] *Potentilla nana* in The Plant List (http://www.theplantlist.org)
[by] *Potentilla sommerfeltii* in The Plant List (http://www.theplantlist.org)
[bz] *Potentilla nivea* ssp. *subquinata* in Rønning (1996)
[ca] It was listed not in The Flora of Svalbard (http://www.svalbardflora.net) but in Rønning (1996)
[cb] *Betula nana* in The Plant List (http://www.theplantlist.org)
[cc] *Empetrum hermaphroditum* in Rønning (1996)
[cd] *Vaccinium gaultherioides* in Rønning (1996)
[ce] *Mertensia maritima* in The Plant List (http://www.theplantlist.org)
[cf] *Gentianella tenella* in Rønning (1996)
[cg] It was listed not in The Flora of Svalbard (http://www.svalbardflora.net) but in Rønning (1996)
[ch] *Arnica alpina* in Dahl (1937) and Neilson (1970); *A. angustifolia* ssp. *angustifolia* in Evju et al. (2010)
[ci] *Erigeron uniflorus* ssp. *eriocephalus* in Evju et al. (2010)
[cj] *Erigeron unalaschkensis* in Dahl (1937)
[ck] *Taraxacum cymbifolium* in Rønning (1996)
[cl] *Campanula rotundifolia* in The Plant List (http://www.theplantlist.org)

Bibliography

Alsos IG, Arnesen G, Sandbakk BE, Elven R (2019) The flora of Svalbard. http://www.svalbardflora.net

Alsos IG, Westergaard K, Lund L, Sandbakk BE (2004) Floraen i Colesdalen, Svalbard (The flora of Colesdalen, Svalbard). Blyttia 62:142–150. (in Norwegian)

Dahl E (1937) On the Vascular Plants of Eastern Svalbard (Skrifter om Svalbrd og Ishavet Nr. 75). Norway's Svalbard and Arctic Ocean Research Survey, Oslo. 50 p

Engelskjøn T (1986) Eco-geographical relations of the Bjørnøya vasular flora, Svalbard. Polar Res 5:79–127

Evju M, Blumentrath S, Hagen D (2010) Nordaust-Svalbard og Søraust-Svalbard naturreservater. Kunnskapsstatus for flora og vegetasjon NINA Rapport 554 (in Norwegian)

Lydersen C, Steen H, Alsos IG (2010) Svalbard. In: Kålås JA, Henriksen S, Skjelseth S, Viken Å (eds) Environmental conditions and impacts for Red List species. Norwegian Biodiversity Information Centre, Trondheim, pp 119–134

Neilson AH (1968) Vascular plants from the northern part of Nordaustlandet, Svalbard. Norsk Polarinstitutt Skrifter Nr. 143. Norsk Polaristitutt, Oslo

Neilson AH (1970) Vascular plants of Edgeøya, Svalbard. Norsk Polarinstitutt Skrifter Nr. 150. Norsk Polaristitutt, Oslo

Rønning O (1996) The flora of Svalbard. Norwegian Polar Institute, Oslo. 184 p

The Plant List (2019) The International Plant Name Index. http://www.theplantlist.org

Common Name – Scientific Name Index

F

Flowering plants

alkali grass (*Puccinellia*), 9, 16, 86

alpine azalea (*Loiseleuria procumbens*), 30

alpine bearberry (*Arctous alpina*
 (= *Arctostaphylos alpina*)), 19, 29, 33

alpine draba (*Draba alpina*), 11, 66, 67

alpine foxtail (*Alopecurus magellanicus*
 (= *A. alpinus*)), 17

alpine sweetgrass (*Hierochloe alpina*), 23

American dwarf birch (*Betula glandulosa*),
 11

American larch (*Larix laricina*), 11

Arctic avens (*Dryas integrifolia*), 9

Arctic bell-heather (*Cassiope tetragona*),
 17, 20, 30–32

Arctic cinquefoil (*Potentilla nana*), 29

Arctic poppy (*Papaver lapponicum*), 11, 28

Arctic raspberry (*Rubus arcticus*), 33

Arctic willow (*Salix arctica*), 9, 11, 17, 19,
 24, 31, 33, 55, 57, 63

Balsam poplar (*Populus balsamifera*), 11

Bellardi bog sedge (*Kobresia myosuroides*),
 17, 19

Bigelow's sedge (*Carex bigelowii*), 17, 19,
 42

black bearberry (*Arctostaphylos
 alpina*), 36

black crowberry (*Empetrum nigrum*), 31, 34

black spruce (*Picea mariana*), 11, 36

bluegrass (*Poa*), 28

blue mountain heather (*Phyllodoce
 caerulea*), 24, 31

bog blueberry (*Vaccinium uliginosum*), 43

bulrush (*Scirpus*), 10

cloudberry (*Rubus chamaemorus*), 11, 33,
 43

common rush (*Juncus*), 16, 42

cottongrass (*Eriophorum*), 10, 24, 34–36,
 86, 87

cottonwood (*Populus*), 20

crowberry (*Empetrum*), 10, 11, 31–33, 36

curly sedge (*Carex rupestris*), 9, 17

diamond-leaf willow (*Salix planifolia*), 19

downy birch (*Betula pubescens*
 (= *B. alba*)), 2

downy willow (*Salix lapponum*), 33

dwarf birch (*Betula nana*), 10, 19, 33, 34,
 36, 43, 60

dwarf fireweed (*Chamerion latifolium*), 17,
 19

dwarf hairgrass (*Deschampsia sukatschewii*
 ssp. *borealis* (= *D. borealis*)), 17

early sandwort (*Minuartia rubella*), 9

felt leaf willow (*Salix alaxensis*), 19

fescue (*Festuca*), 9

Fisher's tundragrass (*Dupontia fisheri*), 35

foxtail (*Alopecurus*), 16

gray willow (*Salix glauca*), 57

green alder (*Alnus viridis*), 11

hare's-tail cottongrass (*Eriophorum
 vaginatum*), 10, 19, 24, 34, 35

Hoppner's sedge (*Carex subspathacea*), 19

ice grass (*Phippsia algida*), 9, 16

Labrador tea (*Ledum palustre*), 10, 11, 30,
 33, 36

Lanate willow (*Salix lanata* ssp.
 richardsonii), 12

Lapland diapensia (*Diapensia lapponica*),
 29, 33

© Springer Nature Switzerland AG 2020
Y. K. Lee, *Arctic Plants of Svalbard*, https://doi.org/10.1007/978-3-030-34560-0

Scientific Name – Common Name Index

F

Flowering plants

Alnus viridis (green alder), 11

Alnus viridis ssp. *crispa* (mountain alder), 33

Alnus viridis ssp. *fruticosa* (Siberian alder), 33

Alopecurus (foxtail), 16

Alopecurus alpinus (alpine foxtail), 17

Alopecurus magellanicus (alpine foxtail), 16, 17

Arctostaphylos (manzanita), 10

Arctostaphylos alpina (alpine bearberry), 19, 29, 33

Arctous alpina (alpine bearberry), 19, 29, 33

Astragalus (milkvetch), 17

Astragalus umbellatus (tundra milk vetch), 17

Betula alba (downy birch), 2

Betula glandulosa (American dwarf birch), 11

Betula nana (dwarf birch), 10, 19, 33, 34, 36, 60

Betula papyrifera (paper birch), 11

Betula pubescens (downy birch), 2

Carex sp. (sedge)

    C. aquatilis ssp. *stans* (water sedge), 42

    C. bigelowii (bigelow's sedge), 17, 19

    C. lugens (spruce muskeg sedge), 12

    C. membranace (membranous sedge), 19

    C. nardina (spike sedge), 9

    C. parallela (parallel sedge), 19

    C. rariflora (looseflower alpine sedge), 35

    C. rupestris (curly sedge), 9, 17

    C. saxatilis (russet sedge), 19

    C. stans (water sedge), 17, 35, 42

    C. subspathacea (Hoppner's sedge), 19

Cassiope tetragona (Arctic bell-heather), 17, 20, 30–32

Cerastium (mouse-ear), 75–78

Chamerion latifolium (dwarf fireweed), 17, 19

Deschampsia borealis (dwarf hairgrass), 17

Deschampsia sukatschewii ssp. *borealis* (dwarf hairgrass), 17

Diapensia lapponica (lapland diapensia), 29

Diapensia lapponica (pincushion plant), 19

Draba alpine (alpine draba), 11

Draba daurica (smooth whitlow-grass), 9

Dryas (mountain avens), 9, 17–19, 24, 28, 29, 45, 58, 61

Dryas integrifolia (Arctic avens), 9

Dupontia (tundra-grass), 16, 35

Dupontia fisheri (Fisher's tundragrass), 35

Empetrum (crowberry), 10, 11, 31, 33, 34, 36

Empetrum nigrum (black crowberry), 31, 34

Eriophorum sp. (cottongrass), 10, 35, 36

    E. angustifolium ssp. *triste* (tall cottongrass), 10, 19, 35

    E. scheuchzeri (white cottongrass), 35

    E. triste (tall cottongrass), 10, 19, 35

    E. vaginatum (hare's-tail cottongrass), 10, 19, 24, 34

Festuca (fescue), 9

Harrimanella hypnoides (moss bell heather), 11, 43

Hierochloe alpina (alpine sweetgrass), 23

© Springer Nature Switzerland AG 2020
Y. K. Lee, *Arctic Plants of Svalbard*, https://doi.org/10.1007/978-3-030-34560-0

Printed in the United States
By Bookmasters